PHYSICAL FITNESS TRAINING

FM 21 - 20

HEADQUARTERS,
DEPARTMENT OF THE ARMY

DISTRIBUTION RESTRICTION: Approved for public release; distribution is unlimited.

CHANGE 1

**HEADQUARTERS
DEPARTMENT OF THE ARMY
Washington, DC, 1 October 1998**

PHYSICAL FITNESS TRAINING

1. Change FM 21-20, 30 September 1992, as follows:

REMOVE OLD PAGES	INSERT NEW PAGES
14-3 to 14-8	14-3 to 14-8.2
14-21 to 14-22	14-21 to 14-22

2. A star (*) marks new or changed material.

3. File this transmittal sheet in front of this publication.

DISTRIBUTION RESTRICTION: proved for public release; distribution is unlimited.

Approved for public release; distribution is unlimited.

FM 21-20
C1

Change

No. 1

HEADQUARTERS
DEPARTMENT OF THE ARMY
Washington, DC, *1 October 1998*

PHYSICAL FITNESS TRAINING

1. FM 21-20, 30 September 1992, is changed as follows:

Remove pages	*Insert pages*
14-3 through 14-8	14-3 through 14-8.2
14-21 and 14-22	14-21 and 14-22

2. A star (*) indicates new or changed material.

3. File this transmittal sheet in front of this publication.

By Order of the Secretary of the Army:

DENNIS J. REIMER
General, United States Army
Chief of Staff

Official:

JOEL B. HUDSON
Administrative Assistant to the
Secretary of the Army

Distribution:
Active Army, Army National Guard of the United States, and U.S. Army Reserve: To be distributed in accordance with Initial Distribution Number (IDN) 110165, requirements for FM 21-20.

Field Manual
No. 21-20

*FM 21-20
Headquarters
Department of the Army
Washington, DC, 30 September 1992

TABLE OF CONTENTS

DISTRIBUTION RESTRICTION: Approved for public release; distribution is unlimited.

This publication supersedes FM 21-20, 28 August 1985.

TABLE OF CONTENTS (CONT.)

Preface

On 5 July 1950, U.S. troops, who were unprepared for the physical demands of war, were sent to battle. The early days of the Korean war were nothing short of disastrous, as U.S. soldiers were routed by a poorly equipped, but well-trained, North Korean People's Army. As American soldiers withdrew, they left behind wounded comrades and valuable equipment their training had not adequately prepared them to carry heavy loads.

The costly lessons learned by Task Force Smith in Korea are as important today as ever. If we fail to prepare our soldiers for their physically demanding wartime tasks, we are guilty of paying lip service to the principle of "Train as you fight." Our physical training programs must do more for our soldiers than just get them ready for the semiannual Army Physical Fitness Test (APFT').

FM 21 -20 is directed at leaders who plan and conduct physical fitness training. It provides guidelines for developing programs which will improve and maintain physical fitness levels for all Army personnel. These programs will help leaders prepare their soldiers to meet the physical demands of war. This manual can also be used as a source book by all soldiers. FM 21-20 was written to conform to the principles outlined in FM 25-100, Training the Force.

The benefits to be derived from a good physical fitness program are many. It can reduce the number of soldiers on profile and sick call, invigorate training, and enhance productivity and mental alertness. A good physical fitness program also promotes team cohesion and combat survivability. It will improve soldiers' combat readiness.

The proponent of this publication is HQ TRADOC. Send comments and recommendations on DA Form 2028 (Recommended Changes to Publications and Blank Forms) directly to Headquarters, US Army Infantry Center, US Army Physical Fitness School (ATZB-PF), Fort Benning, GA31905-5000.

Unless this publication states otherwise, masculine nouns and pronouns do not refer exclusively to men.

CHAPTER 1

Introduction

A soldier's level of physical fitness' has a direct impact on his combat readiness. The many battles in which American troops have fought underscore the important role physical fitness plays on the battlefield. The renewed nationwide interest in fitness has been accompanied by many research studies on the effects of regular participation in sound physical fitness programs. The overwhelming conclusion is that such programs enhance a person's quality of life, improve productivity, and bring about positive physical and mental changes. Not only are physically fit soldiers essential to the Army, they are also more likely to have enjoyable, productive lives.

This chapter provides an overview of fitness. It defines physical fitness, outlines the phases of fitness, and discusses various types of fitness programs and fitness evaluation. Commanders and leaders can use this information to develop intelligent, combat-related, physical fitness programs.

Components of physical fitness include weight control, diet, nutrition, stress management, and spiritual and ethical fitness.

Physical fitness, the emphasis of this manual, is but one component of total fitness. Some of the "others are weight control, diet and nutrition, stress management, dental health, and spiritual and ethical fitness, as well as the avoidance of hypertension, substance abuse, and tobacco use. This manual is primarily concerned with issues relating directly to the development and maintenance of the five components of physical fitness.

The Army's physical fitness training program extends to all branches of the total Army. This includes the USAR and ARNG and encompasses all ages and ranks and both sexes. Its purpose is to physically condition all soldiers throughout their careers beginning with initial entry training (IET). It also includes soldiers with limiting physical profiles who must also participate in physical fitness training.

Commanders and leaders must ensure that all soldiers in their units maintain the highest level of physical fitness in accordance with this manual and with AR 350-15 which prescribes policies, procedures, and responsibilities for the Army physical fitness program.

Leadership Responsibilities

Effective leadership is critical to the success of a good physical training program. Leaders, especially senior leaders, must understand and practice the new Army doctrine of physical fitness. They must be visible and active participants in physical training programs. In short, leaders must lead PT! Their example will emphasize the importance of physical fitness training and will highlight it as a key element of the unit's training mission.

Leaders must emphasize the value of physical training and clearly explain the objectives and benefits of the program. Master Fitness Trainers (MFTs), graduates of a special course taught by the U.S. Army Physical Fitness School, can help commanders do this. However, regardless of the level of technical experience MFTs have, the sole responsibility for good programs rests with leaders at every level.

A poorly designed and executed physical fitness program hurts morale. A good program is well planned and organized, has reasonable yet challenging requirements, and is competitive and progressive. It also has command presence at every level with leaders setting the example for their soldiers.

Leaders should also continually assess their units to determine which specific components of fitness they lack. Once they identify the shortcomings, they should modify their programs to correct the weaknesses.

Leaders should not punish soldiers who fail to perform to standard. Punishment, especially excessive repetitions or additional PT, often does more harm than good. Leaders must plan special training to help soldiers who need it. The application of sound leadership techniques is especially important in bringing physically deficient soldiers up to standard.

'COMMAND FUNCTIONS

Commanders must evaluate the effectiveness of physical fitness training and ensure that it is focused on the unit's missions. They can evaluate its effectiveness by participating in and observing training, relating their fitness programs to the unit's missions, and analyzing individual and unit APFT performance.

Leaders should regularly measure the physical fitness level of every soldier to evaluate his progress and determine the success of the unit's program.

Commanders should assure that qualified leaders supervise and conduct fitness training and use their MFTs, for they have received comprehensive training in this area.

Leaders can learn about fitness training in the following ways:

- Attend the four-week MFT course or one-week Exercise Leaders Course.
- Request a fitness workshop from the Army Physical Fitness School.
- Become familiar with the Army's fitness publications. Important examples include this manual, AR 350-15, and DA Pamphlets 350-15, 350-18, and 350-22.

Commanders must provide adequate facilities and funds to support a program which will improve each soldier's level of physical fitness. They must also be sure that everyone participates, since all individuals, regardless of rank, age, or sex, benefit from regular exercise. In some instances, leaders will need to make special efforts to overcome recurring problems which interfere with regular training.

Leaders must also make special efforts to provide the correct fitness training for soldiers who are physically substandard. "Positive profiling" (DA Form 3349) permits and encourages profiled soldiers to do as much as they can within the limits of their profiles. Those who have been away from the conditioning process because of leave, sickness, injury, or travel may also need special consideration.

Commanders must ensure that the time allotted for physical fitness training is used effectively.

Commanders must ensure that the time alloted for physical fitness training is used effectively.

Training times is wasted by the following:

- Unprepared or unorganized leaders.
- Assignment fo a group which us too large for one leader.
- Insufficient training intensity: it will result in no improvement.
- Rates of progression that are too slow or too fast.
- Extreme faomality that usually emphasizes form over substance. An example would be too many units runs at slow paces or "daily dozen" activities that look impressive but do not result in impovement.
- Inadequate facilities which cause long waiting periods between exercises during a workout and/or between workouts.
- Long rest periods which interfere with progress.

To foster a positive attitude, unit leaders and instructors must be knowledgeable, understanding, and fair, but demanding. They must recognize individual differences and motivate soldiers to put forth their best efforts. However, they must also emphasize training to standard. Attaining a high level of physical fitness cannot be done simply by going through the motions. Hard training is essential.

Commanders must ensure that leaders are familiar with approved techniques, directives, and publications and that they use them. The objective of every commander should be to incorporate the most effective methods of physical training into a balanced program. This program should result in the improved physical fitness of their soldiers and an enhanced ability to perform mission-related tasks.

MFTs can help commanders formulate sound programs that will attain their physical training goals, but commanders must know and apply the doctrine. However, since the responsibility for physical training is the commander's, programs must be based on his own training objectives. These he must develop from his evaluation of the unit's mission-essential task list (METL). Chapter 10 describes the development of the unit's program.

MASTER FITNESS TRAINERS

A Master Fitness Trainer (MFT) is a soldier who has completed either the four-week active-component, two-week reserve-component, or U.S. Military Academy's MFT course work. Although called "masters," MFTs are simply soldiers who know about all aspects of physical fitness training and how soldiers' bodies function. Most importantly, since MFTs are taught to design individual and unit programs, they should be used by commanders as special staff assistants for this purpose.

MFTs can do the following:

- Assess the physical fitness levels of individuals and units.
- Analyze the unit's mission-related tasks and develop sound fitness training programs to support those tasks.
- Train other trainers to conduct sound, safe physical training.
- Understand the structure and function of the human body, especially as it relates to exercise.

Components of Fitness

Physical fitness is the ability to function effectively in physical work, training, and other activities and still have enough energy left over to handle any emergencies which may arise.

The components of physical fitness are as follows:

- Cardiorespiratory (CR) endurance-the efficiency with which the body delivers oxygen and nutrients needed for muscular activity and transports waste products from the cells.
- Muscular strength - the greatest amount of force a muscle or muscle group can exert in a single effort.
- Muscular endurance - the ability of a muscle or muscle group to perform repeated movements with a sub-maximal force for extended periods of times.
- Flexibility-the ability to move the joints (for example, elbow, knee) or any group of joints through an entire, normal range of motion.
- Body composition-the amount of body fat a soldier has in comparison to his total body mass.

Improving the first three components of fitness listed above will have a positive impact on body composition and will result in less fat. Excessive body fat detracts from the other fitness components, reduces performance, detracts from appearance, and negatively affects one's health.

Factors such as speed, agility, muscle power, eye-hand coordination, and eye-foot coordination are classified as components of "motor" fitness. These factors affect a soldier's survivability on the battlefield. Appropriate training can improve these factors within the limits of each soldier's potential. The Army's fitness program seeks to improve or maintain all the components of physical and motor fitness through sound, progressive, mission-specific physical training for individuals and units.

Principles of Exercise

Adherence to certain basic exercise principles is important for developing an effective program. The principles of exercise apply to everyone at all levels of physical training, from the Olympic-caliber athlete to the weekend jogger. They also apply to fitness training for military personnel.

These basic principles of exercise must be followed:

- Regularity. To achieve a training effect, a person must exercise often. One should strive to exercise each of the first four fitness components at least three times a week. Infrequent exercise can do more harm than good. Regularity is also important in resting, sleeping, and following a good diet.
- Progression. The intensity (how hard) and/or duration (how long) of exercise must gradually increase to improve the level of fitness.
- Balance. To be effective, a program should include activities that address all the fitness components, since overemphasizing any one of them may hurt the others.
- Variety. Providing a variety of activities reduces boredom and increases motivation and progress.
- Specificity. Training must be geared toward specific goals. For example, soldiers become better runners if their training emphasizes running. Although swimming is great exercise, it does not improve a 2-mile-run time as much as a running program does.
- Recovery. A hard day of training for a given component of fitness should be followed by an easier training day or rest day for that component and/or muscle group(s) to help permit recovery. Another way to allow recovery is to alternate the muscle groups exercised every other day, especially when training for strength and/or muscle endurance.
- Overload. The work load of each exercise session must exceed the normal demands placed on the body in order to bring about a training effect.

FITT Factors

Factors for a successful training program are Frequency, Intensity, Time, and Type; "FITT".

Certain factors must be part of any fitness training program for it to be successful. These factors are Frequency, Intensity, Time, and Type. The acronym FITT makes it easier to remember them. (See Figure 1- 1.)

FREQUENCY

Army Regulation 350-15 specifies that vigorous physical fitness training will be conducted 3 to 5 times per week. For optimal results, commanders must strive to conduct 5 days of physical training per week. Ideally, at least three exercise sessions for CR fitness, muscle endurance, muscle strength, and flexibility should be performed each week to improve fitness levels. Thus, for example, to obtain maximum gains in muscular strength, soldiers should have at least three strength-training sessions per week. Three physical activity periods a week, however, with only one session each of cardiorespiratory, strength, and flexibility training will not improve any of these three components.

With some planning, a training program for the average soldier can be developed which provides fairly equal emphasis on all the components of physical fitness. The following training program serves as an example.

In the first week, Monday, Wednesday, and Friday are devoted to CR fitness, and Tuesday and Thursday are devoted to muscle endurance and strength. During the second week, the

FITT Factors Applied to Physical Conditioning Program	Cardiorespiratory Endurance	Muscular Strength	Muscular Endurance	Muscular Strength and Muscular Endurance	Flexibility
F — Frequency	3-5 times/week	3 times/week	3-5 times/week	3 times/week	Warm-up and Cool-down: Stretch before and after each exercise session Developmental Stretching: To improve flexibility, stretch 2-3 times/week
I — Intensity	60-90% HRR*	3-7 RM*	12+ RM	8-12 RM	Tension and slight discomfort, NOT PAIN
T — Time	20 minutes or more	The time required to do 3-7 repetitions of each exercise	The time required to do 12+ repetitions of each exercise	The time required to do 8-12 repetitions of each exercise	Warm-up and Cool-down Stretches: 10-15 seconds/stretch Developmental Stretches: 30-60 seconds/stretch
T — Type	Running Swimming Cross-Country Skiing Rowing Bicycling Jumping Rope Walking/Hiking Stair Climbing	Free Weights Resistance Machines Partner-Resisted Exercises Body-Weight Exercises (Pushups/Situps/Pullups/Dips, etc.)			Stretching: Static Passive P.N.F.

* HRR = Heart Rate Reserve   * RM = Repetition Maximum

Figure 1-1

training days are flip-flopped: muscle endurance and strength are trained on Monday, Wednesday, and Friday, and CR fitness is trained on Tuesday and Thursday. Stretching exercises are done in every training session to enhance flexibility. By training continuously in this manner, equal emphasis can be given to developing muscular endurance and strength and to CR fitness while training five days per week.

If the unit's mission requires it, some muscular and some CR training can be done during each daily training session as long as a "hard day/recovery day" approach is used. For example, if a unit has a hard run on Monday, Wednesday, and Friday, it may also choose to run on Tuesday and Thursday. However, on Tuesday and Thursday the intensity and/or distance/time should be reduced to allow recovery. Depending on the time available for each session and the way training sessions are conducted, all components of fitness can be developed using a three-day-per-week schedule. However, a five-day-per-week program is much better than three per week. (See Training Program in Chapter 10.)

Numerous other approaches can be taken when tailoring a fitness program to meet a unit's mission as long as the principles of exercise are not violated. Such programs, when coupled with good nutrition, will help keep soldiers fit to win.

!NTENSITY

Training at the right intensity is the biggest problem in unit programs. The intensity should vary with the type of exercise being done. Exercise for CR development must be strenuous enough to elevate the heart rate to between 60 and 90 percent of the heart rate reserve (HRR). (The calculation of percent HRR is explained in Chapter 2.) Those with low fitness levels should start exercising at a lower training heart rate (THR) of about 60 percent of HRR.

For muscular strength and endurance, intensity refers to the percentage of the maximum resistance that is used for a given exercise. When determining intensity in a strength-training program, it is easier to refer to a "repetition maximum" or "RM." For example, a 1O-RM is the maximum weight that can be correctly lifted 10 times. An 8-12 RM is the weight that can be lifted 8 to 12 times correctly. Doing an exercise "correctly" means moving the weight steadily and with proper form without getting help from other muscle groups by jerking, bending, or twisting the body. For the average person who wants to improve both muscular strength and endurance, an 8-12 RM is best.

The person who wants to concentrate on muscular strength should use weights which let him do three to seven repetitions before his muscles fatigue. Thus, for strength development, the weight used should be a 3-7 RM. On the other hand, the person who wants to concentrate on muscular endurance should use a 12+ RM. When using a 12+ RM as the training intensity, the more repetitions performed per set, over time, the greater will be the improvement in muscular endurance. Conversely, the greater the number of repetitions performed, the smaller will be the gains in strength. For example, a person who regularly trains with a weight which lets him do 100 repetitions per exercise (a 1OO-RM) greatly increases his muscular endurance but minimally improves his muscular strength. (See Chapter 3 for information on resistance training.)

All exercise sessions should include stretching during the warm-up and cool-down. One should stretch so there is slight discomfort, but no pain, when the movement is taken beyond the normal range of motion. (See Chapter 4 for information on stretching.)

All exercises sessions should include stretching during the warm-up and cool-down.

TIME

Like intensity, the time spent exercising depends on the type of exercise being done. At least 20 to 30 continuous minutes of intense exercise must be used in order to improve cardiorespiratory endurance.

For muscular endurance and strength, exercise time equates to the number of repetitions done. For the average soldier, 8 to 12 repetitions with enough resistance to cause muscle failure improves both muscular endurance and strength. As soldiers progress, they will make better strength gains by doing two or three sets of each resistance exercise.

Flexibility exercises or stretches should be held for varying times depending on the objective of the session. For warming-up, such as before a run, each stretch should be held for 10 to 15 seconds. To improve flexibility, it is best to do stretching during the cool-down, with each stretch held for 30 to 60 seconds. If flexibility improvement is a major goal, at least one session per week should be devoted to developing it.

TYPE

Type refers to the kind of exercise performed. When choosing the type, the commander should consider the principle of specificity. For example, to improve his soldiers' levels of CR fitness (the major fitness component in the 2-mile run), he should have them do CR types of exercises. These are discussed in Chapter 2.

Ways to train for muscular strength and endurance are addressed in Chapter 3, while Chapter 4 discusses flexibility. These chapters will help commanders design programs which are tailor-made to their soldiers' needs. The basic rule is that to improve performance, one must practice the particular exercise, activity, or skill he wants to improve. For example, to be good at push-ups, one must do push-ups. No other exercise will improve push-up performance as effectively.

Warm-up and Cool-Down

One must prepare the body before taking part in organized PT, unit sports competition, or vigorous physical activity. A warm-up may help prevent injuries and maximize performance. The warm-up increases the body's internal temperature and the heart rate. The chance of getting injured decreases when the heart, muscles, ligaments, and tendons are properly prepared for exertion. A warm-up should include some running-in-place or slow jogging, stretching, and calisthenics. It should last five to seven minutes and should occur just before the CR or muscular endurance and strength part of the workout. After a proper warm-up, soldiers are ready for a more intense conditioning activity.

Soldiers should cool down properly after each exercise period, regardless of the type of workout. The cool-down serves to gradually slow the heart rate and helps prevent pooling of the blood in the legs and feet. During exercise, the muscles squeeze the blood through the veins. This helps return the blood to the heart. After exercise, however, the muscles relax and no longer do this, and the blood can accumulate in the legs and feet. This can cause a person to faint. A good cool-down will help avoid this possibility.

Soldiers should walk and stretch until their heart rates return to less than 100 beats per minute (BPM) and heavy sweating stops. This usually happens five to seven minutes after the conditioning session.

Phases of Fitness Conditioning

The physical fitness training program is divided into three phases: preparatory, conditioning, and maintenance. The starting phases for different units or individuals vary depending on their age, fitness levels, and previous physical activity.

Young, healthy persons may be able to start with the conditioning phase, while those who have been exercising regularly may already be in the maintenance phase. Factors such as extended field training, leave time, and illness can cause soldiers to drop from a maintenance to a conditioning phase.

Persons who have not been active, especially if they are age 40 or older, should start with the preparatory phase. Many soldiers who fall into this category may be recovering from illness or injury, or they may be just out of high school. Most units will have soldiers in all three phases of training at the same time.

PREPARATORY PHASE

The preparatory phase helps both the cardiorespiratory and muscular systems get used to exercise, preparing the body to handle the conditioning phase. The work load in the beginning must be moderate. Progression from a lower to a higher level of fitness should be achieved by gradual, planned increases in frequency, intensity, and time.

Initially, poorly conditioned soldiers should run, or walk if need be, three times a week at a comfortable pace that elevates their heart rate to about 60 percent HRR for 10 to 15 minutes. Recovery days should be evenly distributed throughout the week, and training should progress slowly. Soldiers should continue at this or an appropriate level until they have no undue fatigue or muscle soreness the day following the exercise. They should then lengthen their exercise session to 16 to 20 minutes and/or elevate their heart rate to about 70 percent HRR by increasing their pace. To be sure their pace is faster, they should run a known distance and try to cover it in less time. Those who feel breathless or whose heart rate rises beyond their training heart rate (THR) while running should resume walking until the heart rate returns to the correct training level. When they can handle an intensity of 70 percent HRR for 20 to 25 minutes, they should be ready for the next phase. Chapter 2 shows how to determine the THR, that is, the right training level during aerobic training.

The preparatory phase for improving muscular endurance and strength through weight training should start easily and progress gradually. Beginning weight trainers should select about 8 to 12 exercises that work all the body's major muscle groups. They should use only very light weights the first week (that is, the first two to three workouts). This is very important, as they must first learn the proper form for each exercise. Light weights will also help minimize muscle soreness and decrease the likelihood of injury to the muscles, joints, and ligaments. During the second week, they should use progressively heavier weights on each resistance exercise. By the end of the second week (four to six workouts), they should know how much weight will let them do 8 to 12 repetitions to muscle failure for each exercise. At this point the conditioning phase begins.

CONDITIONING PHASE

To reach the desired level of fitness, soldiers must increase the amount of exercise and/or the workout intensity as their strength and/or endurance increases.

To improve cardiorespiratory endurance, for example, they must increase the length of time they run. They should start with the preparatory phase and gradually increase the running time by one or two minutes each week until they can run continuously for 20 to 30 minutes. At this point, they can increase the intensity until they reach the desired level of fitness. They should train at least three times a week and take no more than two days between workouts.

For weight trainers, the conditioning phase normally begins during the third week. They should do one set of 8 to 12 repetitions for each of the selected resistance exercises. When they can do more than 12 repetitions of any exercise, they should increase the

Soldiers and units should be encouraged to progress beyond minimum requirements.

weight used on that exercise by about five percent so they can again do only 8 to 12 repetitions. This process continues throughout the conditioning phase. As long as they continue to progress and get stronger while doing only one set of each exercise, it is not necessary for them to do more than one set per exercise. When they stop making progress with one set, they should add another set on those exercises in which progress has slowed. As training progresses, they may want to increase the sets to three to help promote further increases in strength and/or muscle mass.

For maximum benefit, soldiers should do strength training three times a week with 48 hours of rest between workouts for any given muscle group. It helps to periodically do a different type of exercise for a given muscle or muscle group. This adds variety and ensures better strength development.

The conditioning phase ends when a soldier is physically mission-capable and all personal, strength-related goals and unit-fitness goals have been met.

MAINTENANCE PHASE

The maintenance phase sustains the high level of fitness achieved in the conditioning phase. The emphasis here is no longer on progression. A well-designed, 45- to 60-minute workout (including warm-up and cool-down) at the right intensity three times a week is enough to maintain almost any appropriate level of physical fitness. These workouts give soldiers time to stabalize their flexibility, CR endurance, and muscular endurance and strength. However, more frequent training may be needed to reach and maintain peak fitness levels.

Soldiers and units should always be encouraged to progress beyond minimum requirements. Maintaining an optimal level of fitness should become part of every soldier's life-style and should be continued throughout his life.

An effective program uses a variety of activities to develop muscular endurance and strength, CR endurance, and flexibility, and to achieve good body composition. It should also promote the development of coordination as well as basic physical skills. (See Chapter 10 for guidance in constructing a unit program.)

Types of Fitness Programs

The Army has too many types of units with different missions to have one single fitness program for everyone. Therefore, only broad categories of programs and general considerations are covered here. They are classified as unit, individual, and special programs.

UNIT PROGRAMS

Unit programs must support unit missions. A single unit may require several types of programs. Some units, such as infantry companies, have generally the same types of soldiers and MOSS. On the other hand, certain combat--service-support units have many different types of soldiers, each with unique needs. Commanders can develop programs for their own unit by following the principles in this chapter. MFTs know how to help commanders develop programs for their units/soldiers.

Commanders of units composed of both men and women must also understand the physiological differences between the sexes. These are summarized in Appendix A. Although women are able to participate in the same fitness programs as men, they must work harder to perform at the same absolute level of work or exercise. The same holds true for poorly-conditioned soldiers running with well-conditioned soldiers.

To overcome this problem in the case of running, for example, the unit should use ability group runs rather than unit runs. Soldiers in a given ability group will run at a set pace, with groups based on each soldier's most recent 2-mile-run time. Three to six groups per company-sized unit are usually enough. Within each group, each soldier's heart rate while running should be at his own THR. When the run is not intense enough to bring one or more of the soldiers to THR, it is time for those soldiers to move up to the next ability group.

Ability group running does two things more effectively than unit runs: 1) it lets soldiers improve to their highest attainable fitness level; and, 2) it more quickly brings subpar performers up to minimum standards.

It also allows soldiers to train to excel on the APFT which, in turn, helps promotion opportunities. Holding a fit soldier back by making him run at a slow, unit-run pace (normally less than his minimum pace for the 2-mile run on the APFT) hurts his morale and violates the principle of training to challenge.

initial Entry Training (IET)

The training program in basic training (BT) brings soldiers up to the level of physical fitness they need to do their jobs as soldiers. However, the program requires good cadre leadership to ensure that it is appropriate, demanding, and challenging.

Trainees report to active duty at various levels of physical fitness and ability. During basic training they pass through the preparatory into the conditioning phase. During "fill" periods and the first week of training, the focus is on learning and developing the basics of physical fitness.

Training emphasizes progressive conditioning of the whole body. To minimize the risk of injury, exercises must be done properly, and the intensity must progress at an appropriate rate. Special training should be considered for soldiers who fail to maintain the unit's or group's rate of progression. Commanders should evaluate each basic trainee who falls below standard and give him individualized, special assistance to improve his deficiencies.

Additional training should not be used as punishment for a soldier's inability to perform well.

More PT is not necessarily better. Chapter 11 describes how to develop physical training programs in IET units.

Advanced Individual Training (AIT)

Although AIT focuses on technical and MOS-oriented subjects, physical fitness must be emphasized throughout. Most soldiers arriving from basic training are already well into the conditioning phase. Therefore, AIT unit training should focus on preparing soldiers to meet the physical requirements of their initial duty assignments. (See TRADOC Reg. 350-6, Chapter 4.)

Walking, running, and climbing during unit training contribute to physical fitness, but they are not enough. Physical training in AIT requires continued, regular, vigorous exercise which stresses the whole body and addresses all the components of fitness.

By the end of AIT, soldiers must meet APFT standards. With good programs and special training, all healthy AIT graduates should easily be able to demonstrate that they, possess the required level of physical fitness.

By the end of AIT, soldiers must meet APFT standards.

TOE and TDA Units-Active Component

There are many types of units in the Army, and their missions often require different levels of fitness. TOE and TDA units must emphasize attaining and maintaining the fitness level required for the mission.

The unit's standards may exceed the Army's minimums. By regulation (AR 350- 15), the unit's standards can be established by the unit's commander, based on mission requirements.

TOE and TDA Units--Reserve Components

The considerations for the active component also apply to reserve components (RCS). However, since members of RC units cannot participate together in collective physical training on a regular basis, RC unit programs must focus on the individual's fitness responsibilities and efforts. Commanders, however, must still ensure that the unit's fitness level and individual PT programs are maintained. MFTs can give valuable assistance to RC commanders and soldiers.

INDIVIDUAL PROGRAMS

There must be a positive approach to all special fitness training.

Many soldiers are assigned to duty positions that offer little opportunity to participate in collective unit PT programs. Examples are HQDA, MACOM staffs, hospitals, service school staff and faculty, recruiting, and ROTC. In such organizations, commanders must develop leadership environments that encourage and motivate soldiers to accept individual responsibility for their own physical fitness. Fitness requirements are the same for these personnel as for others. Section chiefs and individual soldiers need to use the fundamental principles and techniques outlined in this manual to help them attain and maintain a high level of physical fitness. MFTs can help develop individual fitness programs.

SPECIAL PROGRAMS

The day-to-day unit PT program conducted for most soldiers may not be appropriate for all unit members. Some of them may not be able to exercise at the intensity or duration best suited to their needs.

At least three groups of soldiers may need special PT programs. They are as follows:

- Those who fail the APFT and do not have medical profiles.
- Those who are overweight/overfat according to AR 600-9
- Those who have either permanent or temporary medical profiles.

Leaders must also give special consideration to soldiers who are age 40 or older and to recent arrivals who cannot meet the standards of their new unit.

Special programs must be tailored to each soldier's needs, and trained, knowledgeable leaders should develop and conduct them. This training should be conducted with the unit, If this is impossible, it should at least occur at the same time.

There must be a positive approach to all special fitness training. Soldiers who lack enough upper body strength to do a given number of push-ups or enough stamina to pass the 2-mile run should not be ridiculed. Instead, their shortcomings should be assessed and the information used to develop individualized programs to help them remedy their specific shortcomings. A company-sized unit may have as many as 20 soldiers who need special attention. Only smart planning will produce good programs for all of them.

Commanders must counsel soldiers, explaining that special programs are being developed in their best interests. They must make it clear that standards will be enforced. Next, they should coordinate closely with medical personnel to develop programs that fit the capabilities of soldiers with medical limitations. Each soldier should then begin an individualized program based on his needs.

MFTs know how to assess CR endurance, muscular strength and endurance, flexibility, and body composition. They can also develop thorough, tailor-made programs for all of a unit's special population.

APFT Failures

Although it is not the heart of the Army's physical fitness program, the APFT is the primary instrument for evaluating the fitness level of each soldier. It is structured to assess the muscular endurance of specific muscle groups and the functional capacity of the CR system.

Soldiers with reasonable levels of overall physical fitness should easily pass the APFT. Those whose fitness levels are substandard will fail. Soldiers who fail the APFT must receive special attention. Leaders should analyze their weaknesses and design programs to overcome them. For example, if the soldier is overweight, nutrition and dietary counseling may be needed along with a special exercise program. DA Pam 350-22 outlines several ways to improve a soldier's performance on each of the APFT events.

When trying to improve APFT performances, leaders must ensure that soldiers are not overloaded to the point where the fitness training becomes counterproductive. They should use ability groups for their running program and, in addition to a total-body strength-training program, should include exercises designed for push-up and sit-up improvement. When dealing with special populations, two very important principles are overload and recovery. The quality, not just the quantity, of the workout should be emphasized. Two-a-day sessions, unless designed extremely well, can be counter-productive. More PT is not always better.

Overweight Soldiers

Designers of weight loss and physical training programs for overweight soldiers should remember this: even though exercise is the key to sensible weight loss, reducing the number of calories consumed is equally important. A combination of both actions is best.

The type of exercise the soldier does affects the amount and nature of the weight loss. Both running and walking burn about 100 calories per mile. One pound of fat contains 3,500 calories. Thus, burning one pound of fat through exercise alone requires a great deal of running or walking. On the other hand, weight lost through dieting alone includes the loss of useful muscle tissue. Those who participate in an exercise program that emphasizes the development of strength and muscular endurance, however, can actually increase their muscle mass while losing body fat. These facts help explain why exercise and good dietary practices must be combined.

Unit MFTs can help a soldier determine the specific caloric requirement he needs to safely and successfully lose excess fat. They can devise a sound, individualized plan to arrive at that reduced caloric intake. Likewise, unit MFTs can also develop training programs which will lead to fat loss without the loss of useful muscle tissue.

Generally, overweight soldiers should strive to reduce their fat weight by two pounds per week. When a soldier loses weight, either by diet or exercise or both, a large initial weight loss is not unusual. This may be due to water loss associated with the using up of the body's carbohydrate stores. Although these losses may be encouraging to the soldier, little of this initial weight loss is due to the loss of fat.

Soldiers should be weighed under similar circumstances and at the same time each day. This helps avoid false measurements due to normal fluctuations in their body weight during the day. As a soldier develops muscular endurance and strength, lean muscle mass generally increases. Because muscle weighs more per unit of volume than fat, caution is advised in assessing his progress. Just because a soldier is not losing weight rapidly does not necessarily mean he is not losing fat. In fact, a good fitness program often results in gaining muscle mass while simultaneously losing fat weight. If there is reasonable doubt, his percentage of body fat should be determined.

Soldiers with Profiles

This manual stresses what soldiers can do while on medical profile rather than what they cannot do.

DOD Directive 1308.1 requires that, "Those personnel identified with medically limiting defects shall be placed in a physical fitness program consistent with their limitations as advised by medical authorities."

AR 350-15 states, "For individuals with limiting profiles, commanders will develop physical fitness programs in cooperation with health care personnel."

The Office of the Surgeon General has developed DA Form 3349 to ease the exchange of information between health care personnel and the units. On this form, health care personnel list, along with limitations, those activities that the profiled soldier can do to maintain his fitness level. With this information, the unit should direct profiled soldiers to participate in the activities they can do. (An example of DA Form 3349 is in Appendix B.)

All profiled soldiers should do as much of the regular fitness program as they can, along with substitute activities.

All profiled soldiers should take part in as much of the regular fitness program as they can. Appropriate activities should be substituted to replace those regular activities in which they cannot participate.

Chapter 2 describes some aerobic activities the soldier can do to maintain cardiorespiratory fitness when he cannot run. Chapter 3 shows how to strengthen each body part. Applying this information should allow some strength training to continue even when body parts are injured. The same principle applies to flexibility (Chapter 4).

Medical treatment and rehabilitation should be aimed at restoring the soldier to a suitable level of physical fitness. Such treatment should use appropriate, progressive physical activities with medical or unit supervision.

MFTs can help profiled soldiers by explaining alternative exercises and how to do them safely under the limitations of their profile. MFTs are not, however, trained to diagnose injuries or prescribe rehabilitative exercise programs. This is the domain of qualified medical personnel.

The activity levels of soldiers usually decrease while they are recovering from sickness or injury. As a result, they should pay special attention to their diets to avoid gaining body fat. This guidance becomes more important as soldiers grow older. With medical supervision, proper diet, and the right PT programs, soldiers should be able to overcome their physical profiles and quickly return to their normal routines and fitness levels.

Age as a Factor in Physical Fitness

Soldiers who are age 40 and older represent the Army's senior leadership. On the battlefield, they must lead other soldiers under conditions of severe stress. To meet this challenge and set a good example, these leaders must maintain and demonstrate a high level of physical fitness. Since their normal duties may be stressful but nonphysical, they must take part regularly in a physical fitness program. The need to be physically fit does not decrease with increased age.

People undergo many changes as they grow older. For example, the amount of blood the heart can pump per beat and per minute decreases during maximal exercise, as does the maximum heart rate. This lowers a person's physical ability, and performance suffers. Also, the percent of body weight composed of fat generally increases, while total muscle mass decreases. The result is that muscular strength and endurance, CR endurance, and body composition suffer. A decrease in flexibility also occurs.

Men tend to maintain their peak levels of muscular strength and endurance and CR fitness until age 30. After 30 there is a gradual decline throughout their lives. Women tend to reach their peak in physical capability shortly after puberty and then undergo a progressive decline.

Although a decline in performance normally occurs with aging, those who stay physically active do not have the same rate of decline as those who do not. Decreases in muscular strength and endurance, CR endurance, and flexibility occur to a lesser extent in those who regularly train these fitness components.

Soldiers who are fit at age 40 and continue to exercise show a lesser decrease in many of the physiological functions related to fitness than do those who seldom exercise. A trained 60-year-old, for example, may have the same level of CR fitness as a sedentary 20-year-old. In short, regular exercise can help add life to your years and years to your life.

The assessment phase of a program is especially important for those age 40 and over. However, it is not necessary or desirable to develop special fitness programs for these soldiers. Those who have been exercising regularly may continue to exercise at the same level as they did before reaching age 40. A program based on the principles of exercise and the training concepts in this manual will result in a safe, long-term conditioning program for all soldiers. Only those age 40 and over who have not been exercising regularly may need to start their exercise program at a lower level and progress more slowly than younger soldiers. Years of inactivity and possible abuse of the body cannot be corrected in a few weeks or months.

As of 1 January 1989, soldiers reaching age 40 are no longer required to get clearance from a cardiovascular screening program before taking the APFT. Only a medical profile will exempt them from taking the biannual record APFT. They must, however, have periodic physical examinations in accordance with AR 40-501 and NGR 40-501. These include screening for cardiovascular risk factors.

Evaluation

To evaluate their physical fitness and the effectiveness of their physical fitness training programs, all military personnel are tested biannually using the APFT in accordance with AR 350-15. (Refer to Chapter 14.) However, commanders may evaluate their physical fitness programs more frequently than biannually.

SCORING CATEGORIES

There are two APFT categories of testing for all military personnel Initial Entry Training (IET) and the Army Standard.

IET Standard

The APFT standard for basic training is a minimum of 50 points per event and no less than 150 points overall by the end of basic training. Graduation requirements for AIT and One Station Unit Training (OSUT) require 60 points per event.

Army Standard

All other Army personnel (active and reserve) who are non-IET soldiers must attain the minimum Army standard of at least 60 points per event. To get credit for a record APFT, a medically profiled soldier must, as a minimum, complete the 2-mile run or one of the alternate aerobic events.

Safety is a major consideration when planning and evaluating physical training programs

SAFETY

Safety is a major consideration when planning and evaluating physical training programs. Commanders must ensure that the programs do not place their soldiers at undue risk of injury or accident. They should address the following items:

- Environmental conditions (heat/cold/traction).
- Soldiers' levels of conditioning (low/high/age/sex).
- Facilities (availability/instruction/repair).
- Traffic (routes/procedures/formations).
- Emergency procedures (medical/communication/transport).

The objective of physical training in the Army is to enhance soldiers' abilities to meet the physical demands of war. Any physical training which results in numerous injuries or accidents is detrimental to this goal. As in most training, common sense must prevail. Good, sound physical training should challenge soldiers but should not place them at undue risk nor lead to situations where accidents or injuries are likely to occur.

CHAPTER 2

Cardiorespiratory Fitness

Cardiorespiratory (CR) fitness, sometimes called CR endurance, aerobic fitness, or aerobic capacity, is one of the five basic components of physical fitness. CR fitness is a condition in which the body's cardiovascular (circulatory) and respiratory systems function together, especially during exercise or work, to ensure that adequate oxygen is supplied to the working muscles to produce energy. CR fitness is needed for prolonged, rhythmic use of the body's large muscle groups. A high level of CR fitness permits continuous physical activity without a decline in performance and allows for rapid recovery following fatiguing physical activity.

Activities such as running, road marching, bicycling, swimming, cross-country skiing, rowing, stair climbing, and jumping rope place an extra demand on the cardiovascular and respiratory systems. During exercise, these systems attempt to supply oxygen to the working muscles. Most of this oxygen is used to produce energy for muscular contraction. Any activity that continuously uses large muscle groups for 20 minutes or longer taxes these systems. Because of this, a wide variety of training methods is used to improve cardiorespiratory endurance.

Physiology of Aerobic Training

Aerobic exercise uses oxygen to produce most of the body's energy needs. It also brings into play a fairly complex set of physiological events. To provide enough energy-producing oxygen to the muscles, the following events occur:

- Greater movement of air through the lungs.
- Increased movement of oxygen from the lungs into the blood stream.
- Increased delivery of oxygen-laden blood to the working muscles by the heart's accelerated pumping action.
- Regulation of the blood vessel's size to distribute blood away from inactive tissue to working muscle.
- Greater movemen t of oxygen from the blood into the muscle tissue.
- Accelerated return of veinous blood to the heart.

CR fitness is needed for prolonged, rhythmic use of the body's large muscle groups.

Correctly performed aerobic exercise, over time, causes positive changes in the body's CR system. These changes allow the heart and vascular systems to deliver more oxygen-rich blood to the working muscles during exercise. Also, those muscles regularly used during aerobic exercise undergo positive changes. By using more oxygen, these changes let the muscles make and use more energy during exercise and, as a result, the muscles can work longer and harder.

During maximum aerobic exercise, the trained person has an increased maximum oxygen consumption ($\dot{V}O_2max$). He is better able to process oxygen and fuel and can therefore provide more energy to the working muscles.

$\dot{V}O_2max$, also called aerobic capacity, is the most widely accepted single indicator of one's CR fitness level.

The best way to determine aerobic capacity is to measure it in the laboratory. It is much easier, however, to estimate maximum oxygen uptake by using other methods.

It is possible to determine a soldier's CR fitness level and get an accurate estimate of his aerobic capacity by using his APFT 2-mile-run time. (Appendix F explains how to do this.) Other tests - the bicycle, walk, and step tests - may also be used to estimate one's aerobic capacity and evaluate one's CR fitness level.

In the presence of oxygen, muscle cells produce energy by breaking down carbohydrates and fats. In fact, fats are only used as an energy source when oxygen is present. Hence, aerobic exercise is the best type of activity for attaining and maintaining a low percentage of body fat.

Aerobic exercise is the best type of activity for attaining and maintaining a low percentage of body fat.

A person's maximum aerobic capacity can be modified through physical training. To reach very high levels of aerobic fitness, one must train hard. The best way to improve CR fitness is to participate regularly in a demanding aerobic exercise program.

Many factors can negateively affect one's ability to perform well aerobically. These include the following:

- Age.
- Anemia.
- Carbon monoxide from tobacco smoke or pollution.
- High altitude (reduced oxygen pressure).
- Illness (heart disease).
- Obesity.
- Sedentary life-style.

Any condition that reduces the body's ability to bring in, transport, or use oxygen reduces a person's ability to perform aerobically. Inactivity causes much of the decrease in physical fitness that occurs with increasing age. Some of this decrease in aerobic fitness can be slowed by taking part in a regular exercise program.

Certain medical conditions also impair the transport of oxygen. They include diseases of the lungs, which interfere with breathing, and disabling heart conditions. Another is severe blocking of the arteries which inhibits blood flow to the heart and skeletal muscles.

Smoking can lead to any or all of the above problems and can, in the long and short term, adversely affect one's ability to do aerobic exercise.

FITT Factors

As mentioned in Chapter 1, a person must integrate several factors into any successful fitness training program to improve his fitness level. These factors are summarized by the following words which form the acronym FITT. Frequency, Intensity, Time, and Type. They are described below as they pertain to cardiorespiratory fitness. A warm-up and cool-down should also be part of each workout. Information on warming up and cooling down is given in Chapters 1 and 4.

FREQUENCY

Frequency refers to how often one exercises. It is related to the intensity and duration of the exercise session. Conditioning the CR system can best be accomplished by three adequately intense workouts per week. Soldiers should do these on alternate days. By building up gradually, soldiers can get even greater benefits from working out five times a week. However, leaders should recognize the need for recovery between hard exercise periods and should adjust the training intensity accordingly. They must also be aware of the danger of overtraining and recognize that the risk of injury increases as the intensity and duration of training increases.

INTENSITY

Intensity is related to how hard one exercises. It represents the degree of effort with which one trains and is probably the single most important factor for improving performance. Unfortunately, it is the factor many units ignore.

Changes in CR fitness are directly related to how hard an aerobic exercise is performed. The more energy expended per unit of time, the greater the intensity of the exercise. Significant changes in CR fitness are brought about by sustaining training heart rates in the range of 60 to 90 percent of the heart rate reserve (HRR). Intensities of less than 60 percent HRR are generally inadequate to produce a training effect, and those that exceed 90 percent HRR can be dangerous.

Soldiers should gauge the intensity of their workouts for CR fitness by determining and exercising at their training heart rate (THR). Using the THR method lets them find and prescribe the correct level of intensity during CR exercise. By determining one's maximum heart rate, resting heart rate, and relative conditioning level, an appropriate THR or intensity can be prescribed.

One's ability to monitor the heart rate is the key to success in CR training. (Note: Ability-group running is better than unit running because unit running does not accommodate the individual soldier's THR. For example, some soldiers in a formation may be training at 50 percent HRR and others at 95 percent HRR. As a result, the unit run will be too intense for some and not intense enough for others.)

The heart rate during work or exercise is an excellent indicator of how much effort a person is exerting. Keeping track of the heart rate lets one gauge the intensity of the CR exercise being done. With this information, one can be sure that the intensity is enough to improve his CR fitness level.

Following are two methods for determining training heart rate (THR). The first method, percent maximum heart rate (% MHR), is simpler to use, while the second method, percent heart rate reserve (% HRR), is more accurate. Percent HRR is the recommended technique for determining THR.

Percent MHR Method

With this method, the THR is figured using the estimated maximal heart rate. A soldier determines his estimated maximum heart rate by subtracting his age from 220. Thus, a 20-year-old would have an estimated maximum heart rate (MHR) of 200 beats per minute (220 -20 = 200).

To figure a THR that is 80 percent of the estimated MHR for a 20-year-old soldier in good physical condition, multiply 0.80 times the MHR of 200 beats per minute (BPM). This example is shown below.

FORMULA

% x MHR = THR

CALCULATION

0.80 x 200 BPM = 160 BPM

When using the MHR method, one must compensate for its built-in weakness. A person using this method may exercise at an intensity which is not high enough to cause a training effect. To compensate for this, a person who is in poor shape should exercise at 70 percent of his MHR; if he is in relatively good shape, at 80 percent MHR; and, if he is in excellent shape, at 90 percent MHR.

Intensity is probably the single most important factor for improving performance.

By determining one's maximum heart rate, resting rate, and conditioning level, an appropriate THR can be prescribed.

Percent HRR Method

A more accurate way to calculate THR is the percent HRR method. The range from 60 to 90 percent HRR is the THR range in which people should exercise to improve their CR fitness levels. If a soldier knows his general level of CR fitness, he can determine which percentage of HRR is a good starting point for him. For example, if he is in excellent physical condition, he could start at 85 percent of his HRR; if he is in reasonably good shape, at 70 percent HRR; and, if he is in poor shape, at 60 percent HRR.

Most CR workouts should be conducted with the heart rate between 70 to 75 percent HRR to attain, or maintain, an adequate level of fitness. Soldiers who have reached a high level of fitness may derive more benefit from working at a higher percentage of HRR, particularly if they cannot find more than 20 minutes for CR exercise. Exercising at any lower percentage of HRR does not give the heart, muscles, and lungs an adequate training stimulus.

Before anyone begins aerobic training, he should know his THR (the heart rate at which he needs to exercise to get a training effect).

The example below shows how to figure the THR by using the resting heart rate (RHR) and age to estimate heart rate reserve (HRR). A 20-year-old male soldier in reasonably good physical shape is the example.

STEP 1: Determine the MHR by subtracting the soldier's age from 220.

FORMULA
220 - age = MHR
(GIVEN)

CALCULATION
220 - 20 = 200 BPM

STEP 2: Determine the RHR in beats per minute (BPM) by counting the resting pulse for 30 seconds, and multiply the count by two. A shorter period can be used, but a 30-second count is more accurate. This count should be taken while the soldier is completely relaxed and rested. How to determine heart rate is described below. Next, determine the heart rate reserve (HRR) by subtracting the RHR from the estimated MHR. If the soldier's RHR is 69 BPM, the HRR is calculated as shown here.

FORMULA
MHR - RHR = HRR

CALCULATION
200 BPM - 69 BPM = 131 BPM

STEP 3: Calculate the THR based on 70 percent of HRR (a percentage based on a good level of CR fitness).

FORMULA
(% x HRR) + RHR = THR

CALCULATION
(0.70x131 BPM)+69 BPM=160.7 BPM

As shown, the percentage (70 percent in this example) is converted to the decimal form (0.70) before it is multiplied by the HRR. The result is then added to the resting heart rate (RHR) to get the THR. Thus, the product obtained by multiplying 0.70 and 131 is 91.7. When 91.7 is added to the RHR of 69, a THR of 160.7 results. When the calculations produce a fraction of a heart beat, as in the example, the value is rounded off to the nearest whole number. In this case, 160.7 BPM is rounded off to give a THR of 161 BPM. In summary, a reasonably fit 20-year-old soldier with a resting heart rate of 69 BPM has a training heart rate goal of 161 BPM. To determine the RHR, or to see if one is within the THR during and right after exercise, place the tip of the third finger lightly over one of the carotid arteries in the neck. These arteries are located to the left and right of the Adam's apple. (See Figure 2-1A.) Another convenient spot from which to monitor the pulse is on the radial artery on the wrist just above the base of the thumb. (See Figure 2-1B.) Yet another way is to place the hand over the heart and count the number of heart beats. (See Figure 2-1 C.)

During aerobic exercise, the body will usually have reached a "Steady State" after five minutes of exercise, and the heart rate will have leveled off. At this time, and immediately after exercising, the soldier should monitor his heart rate.

He should count his pulse for 10 seconds, then multiply this by six to get his heart rate for one minute. This will let him determine if his training intensity is high enough to improve his CR fitness level.

For example, use the THR of 161 BPM figured above. During the 10-second period, the soldier should get a count of 27 beats (161/6= 26.83 or 27) if he is exercising at the right intensity. If his pulse rate is below the THR, he must exercise harder to increase his pulse to the THR. If his pulse is above the THR, he should normally exercise at a lower intensity to reduce the pulse rate to the prescribed THR. He should count as accurately as possible, since one missed beat during the 10-second count, multiplied by six, gives an error of six BPM.

SITES TO TAKE THE HEART RATE

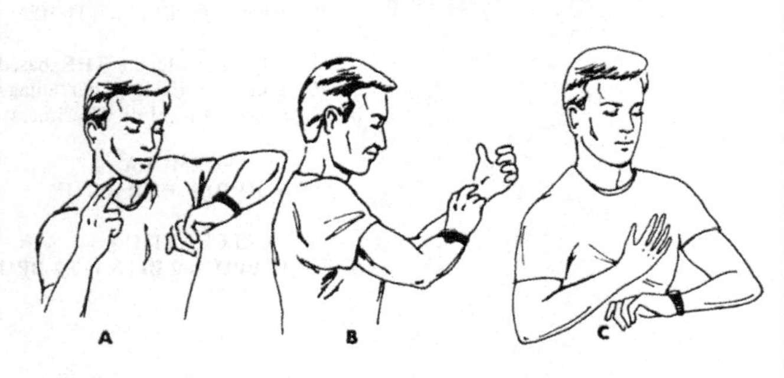

Figure 2-1

A soldier who maintains his THR throughout a 20-30-minute exercise period is doing well and can expect improvement in his CR fitness level.

A soldier who maintains his THR throughout a 20- to 30-minute exercise period is doing well and can expect improvement in his CR fitness level. He should check his exercise and post-exercise pulse rate at least once each workout. If he takes only one pulse check, he should do it five minutes into the workout.

Figure 2-2 is a chart that makes it easy to determine what a soldier's THR should be during a 10-second count. Using this figure, a soldier can easily find his own THR just by knowing his age and general fitness level. For example, a 40-year-old soldier with a low fitness level should, during aerobic exercise, have a THR of 23 beats in 10 seconds. He can determine this from the table by locating his age and then tracking upward until he reaches the percent HRR for his fitness level. Again, those with a low fitness level should work at about 60 percent HRR and those with a good fitness level at **70** percent HRR. Those with a high level of fitness may benefit most by training at 80 to 90 percent HRR.

Another way to gauge exercise intensity is "perceived exertion." This method relies on how difficult the exercise seems to be and is described in Appendix G.

TRAINING HEART RATE

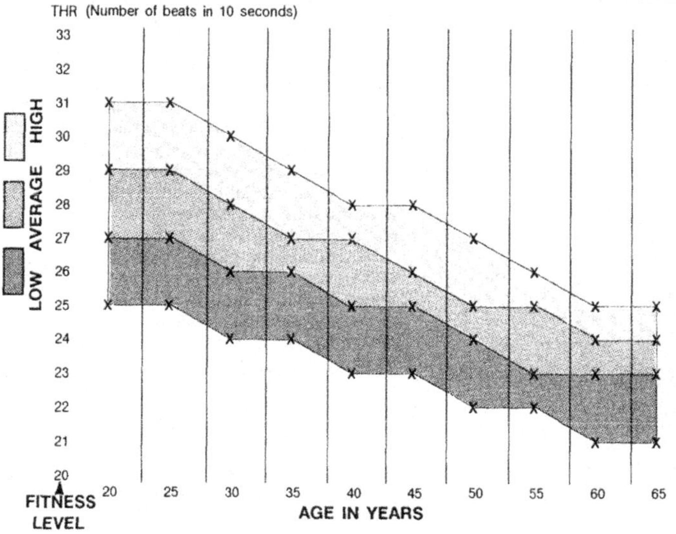

To determine a soldier's THR, locate his age in years. Next, using his general level of fitness (i.e., high, average, or low), find his THR. This will be presented as the number of beats his heart should make during exercise in a 10-second period.

Figure 2-2

TIME

Time, or duration, refers to how long one exercises. It is inversely related to intensity. The more intense the activity, the shorter the time needed to produce or maintain a training effect; the less intense the activity, the longer the required duration. To improve CR fitness, the soldier must train for at least 20 to 30 minutes at his THR.

TYPE

Only aerobic exercises that require breathing in large volumes of air improve CR fitness. Worthwhile aerobic activities must involve the use of large muscle groups and must be rhythmic. They must also be of sufficient duration and intensity (60 to 90 percent HRR). Examples of primary and secondary exercises for improving CR fitness are as follows:

PRIMARY

- Running.
- Rowing.
- Jogging.
- Skiing (cross-country).
- Walking (vigorous).
- Exercising to music.
- Road marching.
- Rope skipping.
- Bicycling (stationary).
- Swimming.
- Bicycling (road/street).
- Stair climbing.

SECONDARY (Done with partners or opponents of equal or greater ability.)

- Racquetball (singles).
- Basketball (full court).
- Handball (singles).
- Tennis (singles).

The primary exercises are more effective than the secondary exercises in producing positive changes in CR fitness. The secondary activities may briefly elevate the heart rate but may not keep it elevated to the THR throughout the entire workout.

Every activity has its advantages and disadvantages. Trainers must weigh these and design programs that fit the unit's needs.

Every activity has its advantages and disadvantages. Trainers must design programs that fit the unit's needs.

Running

Running enables the body to improve the transport of blood and oxygen to the working muscles and brings about positive changes in the muscles' ability to produce energy. Running fits well into any physical training program 'because a training effect can be attained with only three 20-minute workouts per week.

Some soldiers may need instruction to improve their running ability. The following style of running is desired. The head is erect with the body in a straight line or slightly bent forward at the waist. The elbows are bent so the forearms are relaxed and held loosely at waist level. The arms swing naturally from front to rear in straight lines. (Cross-body arm movements waste energy. The faster the run, the faster the arm action.) The toes point straight ahead, and the feet strike on the heel and push off at the big toe.

Besides learning running techniques, soldiers need information on ways to prevent running injuries. The most common injuries associated with PT in the Army result from running and occur to the feet, ankles, knees, and legs. Proper warm-up and cool-down, along with stretching exercises and wearing appropriate clothing and well-fitting running shoes, help prevent injuries. Important information on safety factors and common running injuries is presented in Chapter 13 and Appendix E.

Important information on safety factors and common running injuries is presented in Chapter 13 and Appendix E.

Failure to allow recovery between hard bouts of running cannot only lead to overtraining, but can also be a major cause of injuries. A well-conditioned soldier can run five to six times a week. However, to do this safely, he should do two things: 1) gradually buildup to running that frequently; and, 2) vary the intensity and/or duration of the running sessions to allow recovery between them.

ABILITY GROUP RUNNING

Traditionally, soldiers have run in unit formations at a pace prescribed by the PT leader. Commanders have used unit runs to improve unit cohesion and fitness levels. Unfortunately, too many soldiers are not challenged enough by the intensity or duration of the unit run, and they do not receive a training benefit. For example, take a company that runs at a nine-minute-per-mile pace for two miles. Only soldiers who cannot run two miles in a time faster than 18 minutes will receive a significant training effect. Therefore, in terms of conditioning, most soldiers who can pass the 2-mile-run test are wasting their time and losing the chance to train hard to excel. Ability group running (AGR) is the best way to provide enough intensity so each soldier can improve his own level of CR fitness.

AGR lets soldiers train in groups of near-equal ability. Each group runs at a pace intense enough to produce a training effect for that group and each soldier in it. Leaders should program these runs for specific lengths of time, not miles to be run. This procedure lets more-fit groups run a greater distance than the less-fit groups in the same time period thus enabling every soldier to improve.

The best way to assign soldiers to ability groups is to make a list, in order, of the unit's most recent APFT 2-mile-run times. The number of groups depends on the unit size, number of leaders available to conduct the runs, and range of 2-mile-run times. A company-sized unit broken down into four to six ability groups, each with a leader, is best for aerobic training, For activities like circuits, strength training, and competitive events, smaller groups are easier to work with than one large group.

The best way to assign soldiers to ability groups is to, make a list, in order, of the unit's most recent APFT ***2-mile-run times.***

Because people progress at different rates, soldiers should move to faster groups when they are ready. To help them train at their THR and enhance their confidence, those who have a hard time keeping up with a group should be placed in a slower group. As the unit's fitness level progresses, so should the intensity at which each group exercises. Good leadership will prevent a constant shifting of soldiers between groups due to lack of effort.

AGR is best conducted at the right intensity at least three times a week. As explained, the CR system should not be exercised "hard" on consecutive days. If AGR is used on hard CR-training days, unit runs at lower intensities are good for recovery days. Using this rotation, soldiers can gain the desired benefits of both unit and ability-group runs. The problem comes when units have a limited number of days for PT and there is not enough time for both. In this case, unit runs should seldom, if ever, be used and should be recognized for what they are -- runs to build unit cohesion.

Leaders can use additional methods to achieve both goals. The unit can begin in formation and divide into ability groups at a predetermined release point. The run can also begin with soldiers divided into ability groups which join at a link-up point. Alternately, ability groups can be started over the same route in a stagger, with the slowest group first. Link-ups occur as each faster group overtakes slower groups.

With imagination and planning, AGR will result in more effective training workouts for each soldier. The argument that ability-group running detracts from unit cohesion is invalid. Good leadership and training in all areas promote unit cohesion and team spirit; training that emphasizes form over substance does not.

INTERVAL TRAINING

Interval training also works the cardiorespiratory system. It is an advanced form of exercise training which helps a person significantly improve his fitness level in a relatively short time and increase his running speed.

In interval training, a soldier exercises by running at a pace that is slightly faster than his race pace for short periods of time. This may be faster than the pace he wants to maintain during the next APFT 2-mile run. He does this repeatedly with periods of recovery placed between periods of fast running. In this way, the energy systems used are allowed to recover, and the exerciser can do more fast-paced running in a given workout than if he ran continuously without resting. This type of intermittent training can also be used with activities such as cycling, swimming, bicycling, rowing, and road marching.

The following example illustrates how the proper work-interval times and recovery times can be calculated for interval training so that it can be used to improve a soldier's 2-mile-run performance.

The work-interval time (the speed at which a soldier should run each 440-yard lap) depends on his actual race pace for one mile. If a soldier's actual 1-mile-race time is not known, it can be estimated from his last APFT by taking one half of his 2-mile-run time. Using a 2-mile-run time of 1600 minutes as an example, the pace for an interval training workout is calculated as follows:

Step 1. Determine (or estimate) the actual 1-mile-race pace. The soldier's 2-mile-run time is 16:00 minutes, and his estimated pace for 1 mile is one half of this or 8:00 minutes.

Step 2. Using the time from Step 1, determine the time it took to run 440 yards by dividing the 1-mile-race pace by four. (8:00 minutes/4 = 2:00 minutes per 440 yards.)

Step 3. Subtract one to four seconds from the 440-yard time in Step 2 to find the time each 440-yard lap should be run during an interval training session. (2:00 minutes - 1 to 4 seconds = 1:59 to 1:56.)

Thus, each 440-yard lap should be run in 1 munute, 56 seconds to 1 minute, 59 seconds during interval training based on the soldier's 16:00, 2-mile run time. Recovery periods, twice the length of the work-interval periods. These recovery periods, therefore, will be 3 minutes, 52 seconds long (1:56 + 1:56 = 3:52).

Using the work-interval time for each 440-yard lap from Step 3, the soldier can run six to eight repetitions of 440 yards at a pace of 1 minute, 56 seconds (1:56) for each 440-yard run. This can be done on a 440-yard track (about 400 meters) as follows:

1. Run six to eight 440-yard repetitions with each interval run at a 1:56 pace.
2. Follow each 440-yard run done in 1 minute, 56 secons by an easy jog of 440 yards for recovery. Each 440-yard jog should take twice as much time as the work interval (that is, 3:52). For each second of work, there are two seconds of recovery. Thus, the work-to-rest ratio is 1:2.

440-YARD TIMES FOR INTERVAL TRAINING

1-Mile Time	440- Yard Time
4:45 - 5:00*	1:05 - 1:09*
5:01 - 5:59	1:14 - 1:25
6:00 - 6:59	1:25 - 1:40
7:00 - 7:59	1:41 - 1:55
8:00 - 8:59	1:55 - 2:10
9:00 - 9:59	2:10 - 2.25
10:00 - 10:59	2:25 +

* The slower 1-mile run times correspond to the slower 440-yard times, as do the faster 1-mile times with the faster 440-yard times.

Table 2-1

In Fartlek training, the soldier varies the intensity (speed) of the running throughout the workout.

To help determine the correct time intervals for a wide range of fitness levels, refer to Table 2-1. It shows common 1 -mile times and the corresponding 440-yard times.

Monitoring the heart-rate response during interval training is not as important as making sure that the work intervals are run at the proper speed. Because of the intense nature of interval training, during the work interval the heart rate will generally climb to 85 or 90 percent of HRR. During the recovery interval, the heart rate usually falls to around 120 to 140 beats per minute. Because the heart rate is not the major concern during interval training, monitoring THR and using it as a training guide is not necessary.

As the soldier becomes more conditioned, his recovery is quicker. As a result, he should either shorten the recovery interval (jogging time) or run the work interval a few seconds faster.

After a soldier has reached a good CR fitness level using the THR method, he should be ready for interval training. As with any other new training method, interval training should be introduced into his training program gradually and progressively. At first, he should do it once a week. If he responds well, he may do it twice a week at the most, with at least one recovery day in between. He may also do recovery workouts of easy jogging on off days. It is recommended that interval training be done two times a week only during the last several weeks before an APFT. Also, he should rest the few days before the test by doing no, or very easy, running.

As with any workout, soldiers should start interval workouts with a warm-up and end them with a cool-down.

FARTLEK TRAINING

In Fartlek training, another type of CR training sometimes called speed play, the soldier varies the intensity (speed) of the running during the workout. Instead of running at a constant speed, he starts with veryslow jogging. When ready, he runs hard for a few minutes until he feels the need to slow down. At this time he recovers by jogging at an easy pace. This process of alternating fast and recovery running (both of varying distances) gives the same results as interval training. However, neither the running *nor* recovery interval is timed, and the running is not done on a track. For these reasons, many runners prefer Fartlek training to interval training.

LAST-MAN-UP RUNNING

This type of running, which includes both sprinting and paced running, improves CR endurance and conditions the legs. It consists of 40- to 50-yard sprints at near-maximum effort. This type of running is best done by squads and sections. Each squad leader places the squad in an evenly-spaced, single-file line on a track or a smooth, flat course. During a continuous 2- to 3-mile run of moderate intensity, the squad leader, running in the last position, sprints to the front of the line and becomes the leader. When he reaches the front, he resumes the moderate pace of the whole squad. After he reaches the front, the next soldier, who is now at the rear, immediately sprints to the front. The rest of the soldiers continue to run at a moderate pace. This pattern of sprinting by the last person continues until each soldier has resumed his original position in line. This pattern of sprinting and running is repeated several times during the run. The distance run and number of sprints performed should increase as the soldiers' conditioning improves.

CROSS-COUNTRY RUNNING

Cross-country running conditions the leg muscles and develops CR endurance. It consists of running a certain distance on a course laid out across fields, over hills, through woods, or on any other irregular terrain. It can be used as both a physical conditioning activity and a competitive event. The object is to cover the distance in the shortest time.

The unit is divided into ability groups using 2-mile-run times. Each group starts its run at the same time. This lets the better-conditioned groups run farther and helps ensure that they receive an adequate training stimulus.

The speed and distance can be increased gradually as the soldiers' conditioning improves. At first, the distance should be one mile or less, depending on the terrain and fitness level. It should then be gradually increased to four miles. Cross-country runs have several advantages: they provide variety in physical fitness training, and they can accommodate large numbers of soldiers. Interest can be stimulated by competitive runs after soldiers attain a reasonable level of fitness. These runs may also be combined with other activities such as compass work (orienteering).

Cross-country runs can accommodate large numbers of soldiers.

Road Marches

The road or foot march is one of the best ways to improve and maintain fitness. Road marches are classified as either administrative or tactical, and they can be conducted in garrison or in the field. Soldiers must be able to move quickly, carry a load (rucksack) of equipment, and be physically able to perform their missions after extended marching.

BENEFITS OF ROAD MARCHES

Road marches are an excellent aerobic activity. They also help develop endurance in the muscles of the lower body when soldiers carry a heavy load. Road marches offer several benefits when used as part of a fitness program. They are easy to organize, and large numbers of soldiers can participate. In addition, when done in an intelligent, systematic, and progressive manner, they produce relatively few injuries. Many soldier-related skills can be integrated into road marches. They can also help troops acclimatize to new environments. They help train leaders to develop skills in planning, preparation, and supervision and let leaders make first-hand observations of the soldiers' physical stamina. Because road marches are excellent fitness-training activities, commanders should make them a regular part of their unit's PT program.

Road marches help troops acclimatize to new environments,

TYPES OF MARCHES

The four types of road marches - day, limited visibility, forced, and shuttle - are described below. For more information on marches, see FM 21-18.

Day Marches

Day marches, which fit easily into the daily training plan, are most conducive to developing physical fitness. They are characterized by dispersed formations and ease of control and reconnaissance.

Limited Visibility Marches

Limited visibility marches require more detailed planning and supervision and are harder to control than day marches. Because they move more slowly and are in tighter formations, soldiers may not exercise hard enough to obtain a conditioning effect. Limited visibility marches do have some advantages, however. They protect soldiers from the heat of the day, challenge the ability of NCOS and officers to control their soldiers, and provide secrecy and surprise in tactical situations.

Forced Marches

Soldiers should receive advance notcie before going on a march, to help morale and give them time to prepare.

Forced marches require more than the normal effort in speed and exertion. Although they are excellent conditioners, they may leave soldiers too fatigued to do other required training tasks.

Shuttle Marches

Shuttle marches alternate riding and marching, usually because there are not enough vehicles to carry the entire unit. These marches may be modified and used as fitness activities. A shuttle march can be planned to move troops of various fitness levels from one point to another, with all soldiers arriving at about the same time. Soldiers who have high fitness levels can generally march for longer stretches than those who are less fit.

PLANNING A ROAD MARCH

Any plan to conduct a road march to improve physical fitness should consider the following:

- •Load to be carried.
- •Discipline and supervision.
- •Distance to be marched.
- •Route reconnaissance.
- •Time allotted for movement.
- •Water stops.
- •Present level of fitness.
- •Rest stops.
- •Intensity of the march.
- •Provisions for injuries.
- •Terrain an weather conditions.
- •Safety precautions.

Soldiers should usually receive advance notice before going on a march. This helps morale and gives them time to prepare. The leader should choose an experienced soldier as a pacesetter to lead the march. The pacesetter should carry the same load as the other soldiers and should be of medium height to ensure normal strides. The normal stride for a foot march, according to FM 21-18, is 30 inches. This stride, and a cadence of 106 steps per minute, results in a speed of 4.8 kilometers per hour (kph). When a 10-minute rest is taken each hour, a net speed of 4 kph results.

The pacesetter should keep in mind that ground slope and footing affect stride length. For example, the length decreases when soldiers march up hills or down steep slopes. Normal stride and cadence are maintained easily on moderate, gently rolling terrain unless the footing is muddy, slippery, or rough.

Personal hygiene is important in preventing unnecessary injuries. Before the march, soldiers should cut their toenails short and square them

off, wash and dry their feet, and lightly apply foot powder. They should wear clean, dry socks that fit well and have no holes. Each soldier should take one or more extra pair of socks depending on the length of the march. Soldiers who have had problems with blisters should apply a thin coating of petroleum jelly over susceptible areas. Leaders should check soldiers' boots before the march to make sure that they fit well, are broken in and in good repair, with heels that are even and not worn down.

During halts soldiers should lie down and elevate their feet. If time permits, they should massage their feet, apply powder, and change socks. Stretching for a few minutes before resuming the march may relieve cramps and soreness and help prepare the muscles to continue exercising. To help prevent lower back strain, soldiers should help each other reposition the rucksacks and other loads following rest stops. Soldiers can relieve swollen feet by slightly loosening the laces across their arches.

After marches, soldiers should again care for their feet, wash and dry their socks, and dry their boots.

PROGRAMS TO IMPROVE LOAD-CARRYING ABILITY

The four generalized programs described below can be used to improve the soldiers' load-carrying ability. Each program is based on a different number of days per week available for a PT program.

If only two days are available for PT, both should include exercises for improving CR fitness and muscular endurance and strength. Roughly equal emphasis should be given to each of these fitness components.

If there are only three days available for PT, they should be evenly dispersed throughout the week. Two of the days should stress the development of muscular endurance and strength for the whole body. Although all of the major muscle groups of the body should be trained, emphasis should be placed on the leg (hamstrings and quadriceps), hip (gluteal and hip flexors), low back (spinal erector), and abdominal (rectus abdominis) muscles. These two days should also include brief (2-mile) CR workouts of light to moderate intensity (65 to 75 percent HRR). On the one CR fitness day left, soldiers should take a long distance run (4 to 6 miles) at a moderate pace (70 percent HRR), an interval workout, or an aerobic circuit. They should also do some strength work of light volume and intensity. If four days are available, a road march should be added to the three-day program at least twice monthly. The speed, load, distance, and type of terrain should be varied.

If there are five days, leaders should devote two of them to muscular strength and endurance and two of them to CR fitness. One CR fitness day will use long distance runs; the other can stress more intense workouts including interval work, Fartlek running, or last-man-up running. At least two times per month, the remaining day should include a road march.

Leaders must train and march with their units as much as possible.

Soldiers can usually begin road-march training by carrying a total load equal to 20 percent of their body weight. This includes all clothing and equipment. However, the gender make-up and/or physical condition of a unit may require using a different starting load. Beginning distances should be between five and six miles, and the pace should be at 20 minutes per mile over flat terrain with a hard surface. Gradual increases should be made in speed, load, and distance until soldiers can do the anticipated, worst-case, mission-related scenarios without excessive difficulty or exhaustion. Units should take maintenance marches at least twice a month. Distances should vary from six to eight miles, with loads of 30 to 40 percent of body weight. The pace should be 15 to 20 minutes per mile.

Units should do maintenance marches at least twice a month.

A recent Army study showed that road-march training two times a month and four times a month produced similar improvements in road-marching performance. Thus, twice-monthly road marches appear to produce a favorable improvement in soldiers' abilities to road march if they are supported by a sound PT program (five days per week)

Commanders must establish realistic goals for road marching based on assigned missions. They should also allow newly assigned soldiers and those coming off extended profiles to gradually build up to the unit's fitness level before making them carry maximum loads. This can be done with ability groups.

Road marching should be integrated into all other training. Perhaps the best single way to improve load-earring capacity is to have a regular training program which systematically increases the load and distance. It must also let the soldier regularly practice carrying heavy loads over long distances.

As much as possible, leaders at all levels must train and march with their units. This participation enhances leaders' fitness levels and improves team spirit and confidence, both vital elements in accomplishing difficult and demanding road marches.

Alternate Forms of Aerobic Exercise

Some soldiers cannot run. In such cases, they may use other activities as supplements *or* alternatives. Swimming, bicycling, and cross-country skiing are all excellent endurance exercises and are good substitutes for running. Their drawback is that they require special equipment and facilities that are not always available. As with all exercise, soldiers should start slowly and progress gradually. Those who use non-running activities to such training may not improve running ability. To prepare a soldier for the APFT 2-mile run, there is no substitute for running.

SWIMMING

Swimming is a good alternative to running. Some advantages of swimming include the following:

- o Involvement of all the major muscle groups.
- o Body position that enhances the blood's return to the heart.
- o Partial support of body weight by the water, which minimizes lower body stress in overweight soldiers.

Swimming may be used to improve one's CR fitness level and to maintain and improve CR fitness during recovery from an injury. It is used to supplement running and develop upper body endurance and limited strength. The swimmer should start slowly with a restful stroke. After five minutes, he should stop to check his pulse, compare it with his THR and, if needed, adjust the intensity.

Compared with all the other modes of aerobic exercise presented in this manual (e.g., running, walking, cycling, cross-country skiing, rope jumping, etc.) in swimming alone, one's THR should be lower than while doing the other forms of aerobic exercise. This is because, in swimming, the heart does not beat as fast as when doing the other types of exercise at the same work rate. Thus, in order to effectively train the CR system during swimming, a soldier should set his THR about 10 bpm lower than while running. For example, a soldier whose THR while running is 150 bpm should have a THR of about 140 bpm while swimming. By modifying their THRs in this manner while swimming, soldiers will help to ensure that they are working at the proper intensity.

Non-swimmers can run in waist-to chest-deep water, tread water, and do pool-side kicking for an excellent aerobic workout. They can also do calisthenics in the water. Together these activities combine walking and running with moderate resistance work for the upper body.

For injured soldiers, swimming and aerobic water-training are excellent for improving CR fitness without placing undue stress on injured weight-bearing parts of the body.

CYCLING

Cycling is an excellent exercise for developing CR fitness. Soldiers can bicycle outdoors or on a stationary cycling machine indoors. Road cycling should be intense enough to allow the soldier to reach and maintain THR at least 30 minutes.

Soldiers can alter the cycling intensity by changing gears, adding hill work, and increasing velocity. Distance can also be increased to enhance CR fitness, but the distance covered is not as important as the amount of time spent training at THR. The intensity of a workout can be increased by increasing the resistance against the wheel or increasing the pedaling cadence (number of RPM), For interval training, the soldier can vary the speed and resistance and use periods of active recovery at low speed and/or low resistance.

Cycling should be intense enough to let the soldier reach and maintain THR at least 30 minutes.

For swimming, a soldier should set his THR at about 10 beats per minute lower then when running.

WALKING

Walking is another way to develop cardiorespiratory fitness. It is enjoyable, requires no equipment, and causes few injuries. However, unless walking is done for a long time at the correct intensity, it will not produce any significant CR conditioning.

Sedentary soldiers with a low degree of fitness should begin slowly with 12 minutes of walking at a comfortable pace. The heart rate should be monitored to determine the intensity. The soldier should walk at least four times a week and add two minutes each week to every workout until the duration reaches 45 to 60 minutes per workout. He can increase the intensity by adding hills or stairs.

As the walker's fitness increases, he should walk 45 to 60 minutes at a faster pace. A simple way to increase walking speed is to carry the arms the same way as in running. With this technique the soldier has a shorter arm swing and takes steps at a faster rate. Swinging the arms faster to increase the pace is a modified form of race walking (power walking) which allows for more upper-body work. This method may also be used during speed marches. After about three months, even the most unfit soldiers should reach a level of conditioning that lets them move into a running program.

CROSS-COUNTRY SKIING

Cross-country skiing requires vigorous movement of the arms and legs, developing muscular and CR endurance.

Cross-country or Nordic skiing is another excellent alternative to the usual CR activities. It requires vigorous movement of the arms and legs which develops muscular and CR endurance and coordination. Some of the highest levels of aerobic fitness ever measured have been found in cross-country skiers.

Although some regions lack snow, one form or another of cross-country skiing can be done almost anywhere--on country roads, golf courses, open fields, and in parks and forests.

Cross-country skiing is easy to learn. The action is similar to that used in brisk walking, and the intensity may be varied as in running. The work load is determined by the difficulty of terrain, the pace, and the frequency and duration of rest periods. Equipment is reasonably priced, with skis, boots, and poles often obtainable from the outdoor recreation services.

ROPE SKIPPING

Rope skipping is also a good exercise for developing CR fitness. It requires little equipment, is easily learned, may be done almost anywhere, and is not affected by weather. Some runners use it as a substitute for running during bad weather.

A beginner should select a jump rope that, when doubled and stood on, reaches to the armpits. Weighted handles or ropes may be used by better-conditioned soldiers to improve upper body strength. Rope skippers should begin with five minutes of jumping rope and then monitor their heart rate. They should attain and maintain their THR to ensure a training effect, and the time spent jumping should be increased as the fitness level improves.

Rope jumping, however, may be stressful to the lower extremities and therefore should be limited to no more than three times a week. Soldiers should skip rope on a cushioned surface such as a mat or carpet and should wear cushioned shoes.

HANDBALL AND RACQUET SPORTS

Handball and the racquet sports (tennis, squash, and racquetball) involve bursts of intense activity for short periods. They do not provide the same degree of aerobic training as exercises of longer duration done at lower intensities. However, these sports are good supplements and can provide excellent aerobic benefits depending on the skill of the players. If played vigorously each day, they may be an adequate substitute for low-level aerobic training. Because running increases endurance, it helps improve performance in racket sports, but the reverse is not necessarily true.

EXERCISE TO MUSIC

Aerobic exercise done to music is another excellent alternative to running. It is a motivating, challenging activity that combines exercise and rhythmic movements. There is no prerequisite skill, and it can be totally individualized to every fitness level by varying the frequency, intensity, and duration. One can move to various tempos while jogging or doing jumping jacks, hops, jumps, or many other calisthenics.

Workouts can be done in a small space by diverse groups of varying fitness levels. Heart rates should be taken during the conditioning phase to be sure the workout is sufficiently intense. If strengthening exercises are included, the workout addresses every component of fitness. Holding relatively light dumbbells during the workout is one way to increase the intensity for the upper body and improve muscular endurance. Warm-up and cool-down stretches should be included in the aerobic workout.

CHAPTER 3

Muscular Endurance and Strength

On today's battlefield, in addition to cardiorespiratory fitness, soldiers need a high level of muscular endurance and strength. In a single day they may carry injured comrades, move equipment, lift heavy tank or artillery rounds, push stalled vehicles, or do many other strength-related tasks. For example, based on computer-generated scenarios of an invasion of Western Europe, artillerymen may have to load from 300 to 500, 155mm-howitzer rounds (95-lb rounds) while moving from 6 to 10 times each day over 8 to 12 days. Infantrymen may need to carry loads exceeding 100 pounds over great distances, while supporting units will deploy and displace many times. Indeed, survival on the battlefield may, in large part, depend on the muscular endurance and strength of the individual soldier.

Muscular Fitness

Muscular fitness has two components: muscular strength and muscular endurance.

Muscular strength is the greatest amount of force a muscle or muscle group can exert in a single effort.

Muscular endurance is the ability of a muscle or muscle group to do repeated contractions against a less-than-maximum resistance for a given time.

Although muscular endurance and strength are separate fitness components, they are closely related. Progressively working against resistance will produce gains in both of these components.

Muscular Contractions

Isometric, isotonic, and isokinetic muscular endurance and strength are best produced by regularly doing each specific kind of contraction. They are described here.

Isometric contraction produces contraction but no movement, as when pushing against a wall. Force is produced with no change in the angle of the joint.

Isotonic contraction causes a joint to move through a range of motion against a constant resistance. Common examples are push-ups, sit-ups, and the lifting of weights.

Isokinetic contraction causes the angle at the joint to change at a constant rate, for example, at 180 degrees per second. To achieve a constant speed of movement, the load or resistance must change at different joint angles to counter the varying forces produced by the muscle(s) at different angles. This requires the use of isokinetic machines. There are other resistance-training machines which, while not precisely controlling the speed of movement, affect it by varying the resistance throughout the range of motion. Some of these devices are classified as pseudo-isokinetic and some as variable-resistance machines.

Isotonic and isokinetic contractions have two specific phases - the concentric or "positive" phase and the eccentric or "negative" phase. In the concentric phase (shortening) the muscle contracts, while in the eccentric phase (elongation) the muscle returns to its normal length. For example, on the upward phase of the biceps curl, the biceps are shortening. This is a concentric (positive) contraction. During the lowering phase of the curl the biceps are lengthening. This is an eccentric (negative) contraction.

A muscle can control more weight in the eccentric phase of contraction than it can lift concentrically. As a result, the muscle may be able to handle more of an overload eccentrically. This greater overload, in return, may produce greater strength gains.

The nature of the eccentric contraction, however, makes the muscle and connective tissue more susceptible to damage, so there is more muscle soreness following eccentric work.

When a muscle is overloaded, whether by isometric, isotonic, or isokinetic contractions, it adapts by becoming stronger. Each type of contraction has advantages and disadvantages, and each will result in strength gains if done properly.

The above descriptions are more important to those who assess strength than to average people trying to develop strength and endurance. Actually, a properly designed weight training program with free weights or resistance machines will result in improvements in all three of these categories.

Principles of Muscular Training

To have a good exercise program, the seven principles of exercise, described in Chapter 1, must be applied to all muscular endurance and strength training. These principles are overload, progression, specificity, regularity, recovery, balance, and variety.

OVERLOAD

The overload principle is the basis for all exercise training programs. For a muscle to increase in strength, the workload to which it is subjected during exercise must be increased beyond what it normally experiences. In other words, the muscle must be overloaded. Muscles adapt to increased workloads by becoming larger and stronger and by developing greater endurance.

To understand the principle of overload, it is important to know the following strength-training terms:

- Full range of motion. To obtain optimal gains, the overload must be applied thoughout the full range of motion. Exercise a joint and its associated muscles through its complete range starting from the prestretched position (stretched past the relaxed position) and ending in a fully contratcted position. This is crucial to strength development.
- Repetition. When an exercise has progressed through one complete range of motion and back to the beginning, one repetition has been completed.
- One-repetition maximum (1-RM). This is a repetition performed against the greatest possible resistance (the maximum weight a person can lift one time). A 10-RM is the maximum weight one can lift correctly 10 times. Similarly, an 8-12 RM is that weight which allows a person to do from 8 to 12 correct repetitions. The intensity for muscular endurance and strength training is often expressed as a percentage of. the 1-RM.
- Set. This is a series of repetitions done without rest.
- Muscle Failure. This is the inability of a person to do another correct repetition in a set.

The minimum resistance needed to obtain strength gains is 50 percent of the 1 -RM. However, to achieve enough overload, programs are designed to require sets with 70 to 80 percent of one's 1 -RM. (For example, if a soldier's 1 -RM is 200 pounds, multiply 200 pounds by 70 percent [200 X 0.70 = 140 pounds] to get 70 percent of the 1 -RM.)

When a muscle is overloaded by isometric, isotonic, or isokinetic contractions, it adapts by becoming stronger.

A better and easier method is the repetition maximum (RM) method. The exerciser finds and uses that weight which lets him do the correct number of repetitions. For example, to develop both muscle endurance and strength, a soldier should choose a weight for each exercise which lets him do 8 to 12 repetitions to muscle failure. (See Figure 3-1.) The weight should be heavy enough so that, after doing from 8 to 12 repetitions, he momentarily cannot correctly do another repetition. This weight is the 8-12 RM for that exercise.

FITT Factors Applied to Conditioning Programs for Muscular Endurance and/or Strength		
Muscular Strength	**Muscular Endurance**	**Muscular Strength and Muscular Endurance**
3 times/week	3-5 times/week	3 times/week
3-7 RM*	12+ RM	8-12 RM
The time required to do 3-7 repetitions of each resistance exercise	The time required to do 12+ repetitions of each resistance exercise	The time required to do 8-12 repetitions of each resistance exercise
Free Weights Resistance Machines Partner-Resisted Exercises Body-Weight Exercises (Push-ups/Sit-ups/Pull-ups/Dips, etc.)		
* RM = Repetition Maximum		

Figure 3-1

MUSCULAR ENDURANCE/ STRENGTH DEVELOPMENT

To develop muscle strength, the weight selected should be heavier and the RM will also be different. For example, the soldier should find that weight for each exercise which lets him do 3 to 7 repetitions correctly. This weight is the 3-7 RM for that exercise. Although the greatest improvements seem to come from resistances of about 6-RM, an effective range is a 3-7 RM. The weight should be heavy enough so that an eighth repetition would be impossible because of muscle fatigue.

The weight should also not be too heavy. If one cannot do at least three repetitions of an exercise, the resistance is too great and should be reduced. Soldiers who are just beginning a resistance-training program should not start with heavy weights. They should first build an adequate foundation by training with an 8-12 RM or a 12+ RM.

To develop muscular endurance, the soldier should choose a resistance that lets him do more than 12 repetitions of a given exercise. This is his 12+ repetition maximum (12+ RM). With continued training, the greater the number of repetitions per set, the greater will be the improvement in muscle endurance and the smaller the gains in strength. For example, when a soldier trains with a 25-RM weight, gains in muscular endurance will be greater than when using a 15-RM weight, but the gain in strength will not be as great. To optimize a soldier's performance, his RM should be determined from an analysis of the critical tasks of his mission. However, most soldiers will benefit most from a resistance-training program with an 8-12 RM.

Whichever RM range is selected, the soldier must always strive to overload his muscles. The key to overloading a muscle is to make that muscle exercise harder than it normally does.

An overload may be achieved by any of the following methods:

- Increasing the resistance.
- Increasing the number of repetitions per set.
- Increasing the number of sets.
- Reducing the rest time between sets.
- Increasing the speed of movement in the concentric phase.
  (Good form is more important than the speed of movement.)
- Using any combination of the above.

PROGRESSION

When an overload is applied to a muscle, it adapts by becoming stronger and/or by improving its endurance. Usually significant increases in strength can be made in three to four weeks of proper training depending on the individual. If the workload is not progressively increased to keep pace with newly won strength, there will be no further gains. When a soldier can correctly do the upper limit of repetitions for the set without reaching muscle failure, it is usually time to increase the resistance. For most soldiers, this upper limit should be 12 repetitions.

For example, if his plan is to do 12 repetitions in the bench press, the soldier starts with a weight that causes muscle failure at between 8 and 12 repetitions (8- 12 RM). He should continue with that weight until he can do **12** repetitions correctly. He then should increase the weight by about 5 percent but no more than 10 percent. In a multi-set routine, if his goal is to do three sets of eight repetitions of an exercise, he starts with a weight that causes muscle failure before he completes the eighth repetition in one or more of the sets. He continues to work with that weight until he can complete all eight repetitions in each set, then increases the resistance by no more than 10 percent.

SPECIFICITY

A resistance-training program should provide resistance to the specific muscle groups that need to be strengthened. These groups can be identified by doing a simple assessment. The soldier slowly does work-related movements he wants to improve and, at the same time, he feels the muscles on each side of the joints where motion occurs. Those muscles that are contracting or becoming tense during the movement are the muscle groups involved. If the soldier's performance of a task is not adequate or if he wishes to improve, strength training for the identified muscle(s) will be beneficial. To improve his muscular endurance and strength. in a given task, the soldier must do resistance movements that are as similar as possible to those of doing the task. In this way, he ensures maximum carryover value to his soldiering tasks.

REGULARITY

Exercise must be done regularly to produce a training effect. Sporadic exercise may do more harm than good. Soldiers can maintain a moderate level of strength by doing proper strength workouts only once a week, but three workouts per week are best for optimal gains. The principle of regularity also applies to the exercises for individual muscle groups. A soldier can work out three times a week, but when different muscle groups are exercised at each workout, the principle of regularity is violated and gains in strength are minimal.

Exercise must be done regularly to produce a training effect.

RECOVERY

There should be at least a 48-hour recovery period between workouts for the same muscle group.

Consecutive days of hard resistance training for the same muscle group can be detrimental. The muscles must be allowed sufficient recovery time to adapt. Strength training can be done every day only if the exercised muscle groups are rotated, so that the same muscle or muscle group is not exercised on consecutive days. There should be at least a 48-hour recovery period between workouts for the same muscle groups. For example, the legs can be trained with weights on Monday, Wednesday, and Friday and the upper body muscles on Tuesday, Thursday, and Saturday.

Recovery is also important within a workout. The recovery time between different exercises and sets depends, in part, on the intensity of the workout. Normally, the recovery time between sets should be 30 to 180 seconds.

BALANCE

It is important to include exercises that work all the major muscle groups in both the upper and lower body.

When developing a strength training program, it is important to include exercises that work all the major muscle groups in both the upper and lower body. One should not work just the upper body, thinking that running will strengthen the legs.

Most muscles are organized into opposing pairs. Activating one muscle results in a pulling motion, while activating the opposing muscle results in the opposite, or pushing, movement. When planning a training session, it is best to follow a pushing exercise with a pulling exercise which results in movement at the same joint(s). For example, follow an overhead press with a lat pull-down exercise. This technique helps ensure good strength balance between opposing muscle groups which may, in turn, reduce the risk of injury. Sequence the program to exercise the larger muscle groups first, then the smaller muscles. For example, the lat pull-down stresses both the larger latissimus dorsi muscle of the back and the smaller biceps muscles of the arm. If curls are done first, the smaller muscle group will be exhausted and too weak to handle the resistance needed for the lat pull-down. As a result, the soldier cannot do as many repetitions with as much weight as he normally could in the lat pull-down. The latissimus dorsi muscles will not be overloaded and, as a result, they may not benefit very much from the workout.

The best sequence to follow for a total-body strength workout is to first exercise the muscles of the hips and legs, followed by the muscles of the upper back and chest, then the arms, abdominal, low back, and neck. As long as all muscle groups are exercised at the proper intensity, improvement will occur.

VARIETY

A major challenge for all fitness training programs is maintaining enthusiasm and interest. A poorly designed strength- training program can be very boring. Using different equipment, changing the exercises, and altering the volume and intensity are good ways to add variety, and they may also produce better results. The soldier should periodically substitute different exercises for a given muscle group(s). For example, he can do squats with a barbell instead of leg presses on a weight machine. Also, for variety or due to necessity (for example, when in the field), he can switch to partner-resisted exercises or another form of resistance training. However, frequent wholesale changes should be avoided as soldiers may become frustrated if they do not have enough time to adapt or to see improvements in strength.

Workout Techniques

Workouts for improving muscular endurance or strength must follow the principles just described. There are also other factors to consider, namely, safety, exercise selection, and phases of conditioning.

SAFETY FACTORS

Major causes of injury when strength training are improper lifting techniques combined with lifting weights that are too heavy. Each soldier must understand how to do each lift correctly before he starts his strength training program.

The soldier should always do weight training with a partner, or spotter, who can observe his performance as he exercises. To ensure safety and the best results, both should know how to use the equipment and the proper spotting technique for each exercise.

A natural tendency in strength training is to see how much weight one can lift. Lifting too much weight forces a compromise in form and may lead to injury. All weights should be selected so that proper form can be maintained for the appropriate number of repetitions.

Correct breathing is another safety factor in strength training. Breathing should be constant during exercise. The soldier should never hold his breath, as this can cause dizziness and even loss of consciousness. As a general rule, one should exhale during the positive (concentric) phase of contraction as the weight or weight stack moves away from the floor, and inhale during the negative (eccentric) phase as the weight returns toward the floor.

EXERCISE SELECTION

When beginning a resistance-training program, the soldier should choose about 8 to 16 exercises that work all of the body's major muscle groups. Usually eight well-chosen exercises will serve as a good starting point. They should include those for the muscles of the leg, low back, shoulders, and so forth. The soldier should choose exercises that work several muscle groups and try to avoid those that isolate single muscle groups. This will help him train a greater number of muscles in a given time. For example, doing lat pull-downs on the "lat machine" works the latissimus dorsi of the back and the biceps muscles of the upper arm. On the other hand, an exercise like concentration curls for the biceps muscles of the upper arm, although an effective exercise, only works the arm flexor muscles. Also, the concentration curl requires twice as much time as lat pull-downs because only one arm is worked at a time.

Perhaps a simpler way to select an exercise is to determine the number of joints in the body where movement occurs during a repetition. For most people, especially beginners, most of the exercises in the program should be "multi-joint" exercises. The exercise should provide movement at more than one joint. For example, the pull-down exercise produces motion at both the shoulder and elbow joints. The concentration curl, however, only involves the elbow joint.

PHASES OF CONDITIONING

There are three phases of conditioning: preparatory, conditioning, and maintenance. These are also described in Chapter 1.

Preparatory Phase

The soldier should use very light weights during the first week (the preparatory phase) which includes the first two to three workouts. This is very important, because the beginner must concentrate at first on learning

The three phases of conditioning are preparatory, conditioning, and maintenance.

the proper form for each exercise. Using light weights also helps minimize muscle soreness and decreases the likelihood of injury to the muscles, joints, and ligaments. During the second week, he should use progressively heavier weights. By the end of the second week (4 to 6 workouts), he should know how much weight on each exercise will allow him to do 8 to 12 repetitions to muscle failure. If he can do only seven repetitions of an exercise, the weight must be reduced; if he can do more than 12, the weight should be increased.

Conditioning Phase

The third week is normally the start of the conditioning phase for the beginning weight trainer. During this phase, the soldier should increase the amount of weight used and/or the intensity of the workout as his muscular strength and/or endurance increases. He should do one set of 8 to 12 repetitions for each of the heavy-resistance exercises. When he can do more than 12 repetitions of any exercise, he should increase the weight until he can again do only 8 to 12 repetitions. This usually involves an increase in weight of about five percent. This process continues indefinitely. As long as he continues to progress and get stronger, he does not need to do more than one set per exercise. If he stops making progress with one set of 8 to 12 repetitions per exercise, he may benefit from adding another set of 8 to 12 repetitions on those exercises in which progress has slowed. As time goes on and he progresses, he may increase the number to three sets of an exercise to get even further gains in strength and/or muscle mass. Three sets per exercise is the maximum most soldiers will ever need to do.

Maintenance Phase

Once the soldier reaches a high level of fitness, the maintenance phase is used to maintain that level. The emphasis in this phase is no longer on progression but on retention. Although training three times a week for muscle endurance and strength gives the best results, one can maintain them by training the major muscle groups properly one or two times a week. More frequent training, however, is required to reach and maintain peak fitness levels. Maintaining the optimal level of fitness should become part of each soldier's life-style and training routine. The maintenance phase should be continued throughout his career and, ideally, throughout his life.

As with aerobic training, the soldier should do strength training three times a week and should allow at least 48 hours of rest from resistance training between workouts for any given muscle group.

TIMED SETS

Timed sets refers to a method of physical training in which as many repetitions as possible of a given exercise are performed in a specified period of time. After an appropriate period of rest, a second, third, and so on, set of that exercise is done in an equal or lesser time period. The exercise period, recovery period, and the number of sets done should be selected to make sure that an overload of the involved muscle groups occurs.

The use of timed sets, unlike exercises performed in cadence or for a specific number of repetitions, helps to ensure that each soldier does as many repetitions of an exercise as possible within a period of time. It does not hold back the more capable performer by restricting the number of repetitions he may do. Instead, soldiers at all levels of fitness can individually do the number of repetitions they are capable of and thereby be sure they obtain an adequate training stimulus.

In this FM, timed sets will be applied to improving soldier's sit-up and push-up performance. (See Figures 3-2 and 3-3.) Many different but equally valid approaches can be taken when using timed sets to improve push-up and sit-up performance. Below, several of these will be given.

It should first be stated that improving sit-up and push-up performance, although important for the APFT, should not be the main goal of an Army physical training program. It must be to develop an optimal level of physical fitness which will help soldiers carry out their mission during combat. Thus, when a soldier performs a workout geared to develop muscle endurance and strength, the goal should be to develop sufficient strength and/or muscle endurance in all the muscle groups he will be called upon to use as he performs his mission. To meet this goal, and to be assured that all emergencies can be met, a training regimen which exercises all the body's major muscle groups must be developed and followed. Thus, as a general rule, a muscle endurance or strength training workout should not be designed to work exclusively, or give priority to, those muscle groups worked by the sit-up or push-up event.

For this reason, the best procedure to follow when doing a resistance exercise is as follows. First, perform a workout to strengthen all of the body's major muscles. Then, do timed sets to improve push-up and sit-up performance. Following this sequence ensures that all major muscles are worked. At the same time, it reduces the amount of time and work that must be devoted to push-ups and sit-ups. This is because the muscles worked by those two exercises will already be pre-exhausted.

The manner in which timed sets for push-ups and sit-ups are conducted should occasionally be varied. This ensures continued gains and minimizes boredom. This having been said, here is a very time-efficient way of conducting push-up/sit-up improvement. Alternate timed sets of push-ups and timed sets of sit-ups with little or no time between sets allowed for recovery. In this way, the muscle groups used by the push-up can recover while the muscles used in the sit-up are exercised, and vice versa. The following is an example of this type of approach:

TIMED SETS

SET NO.	ACTIVITY	TIME PERIOD	REST INTERVAL
1	Push-ups	45 seconds	0
2	Sit-ups	45 seconds	0
3	Push-ups	30 seconds	0
4	Sit-ups	30 seconds	0
5	Push-ups	30 seconds	0
6	Sit-ups	30 seconds	0

Figure 3-2

If all soldiers exercise at the same time, the above activity can be finished in about 3.5 minutes. As the soldiers' levels of fitness improve, the difficulty of the activity can be increased. This is done by lengthening the time period of any or all timed sets, by decreasing any rest period between timed sets, by increasing the number of timed sets performed, or by any combination of these.

To add variety and increase the overall effectiveness of the activity, different types of push-ups (regular, feet-elevated, wide-hand, close-hand, and so forth) and sit-ups (regular, abdominal twists, abdominal curls, and so forth) can be done. When performing this type of workout, pay attention to how the soldiers are responding, and make adjustments accordingly. For example, the times listed in the chart above may prove to be too long or too short for some soldiers. In the same way, because of the nature of the sit-up, it may become apparent that some soldiers can benefit by taking slightly more time for timed sets of sit-ups than for push-ups.

When using timed sets for push-up and sit-up improvement, soldiers can also perform all sets of one exercise before doing the other. For example, several timed sets of push-ups can be done followed by several sets of sit-ups, or vice versa. With this approach, rest intervals must be placed between timed sets. The following example can be done after the regular strength workout and is reasonable starting routine for most soldiers.

During a timed set of push-ups, a soldier may reach temporary muscle failure at any time before the set is over. If this happens, he should immediately drop to his knees and continue doing modified push-ups on his knees.

Finally, as in any endeavor, soldiers must set goals for themselves. This applies when doing each timed set and when planning for their next and future APFTs.

Major Muscle Groups

In designing a workout it is important to know the major muscle groups, where they are located, and their primary action. (See Figure 3-4.)

To ensure a good, balanced workout, one must do at least one set of exercises for each of the major muscle groups.

TIMED SETS

SET NO.	ACTIVITY	TIME PERIOD	REST INTERVAL
1	Regular Push-ups	30 seconds	30 seconds
2	Wide-hand Push-ups	30 seconds	30 seconds
3	Close-hand Push-ups	30 seconds	30 seconds
4	Regular Push-ups	20 seconds	30 seconds
5	Regular Push-ups done on knees	30 seconds	30 seconds
6	Regular Sit-ups	60 seconds	30 seconds
7	Abdominal Twists	40 seconds	30 seconds
8	Curl-ups	30 seconds	30 seconds
9	Abdominal Crunches	30 seconds	End

Figure 3-3

MAJOR MUSCLE GROUPS

The Major Skeletal Muscles of the Human Body

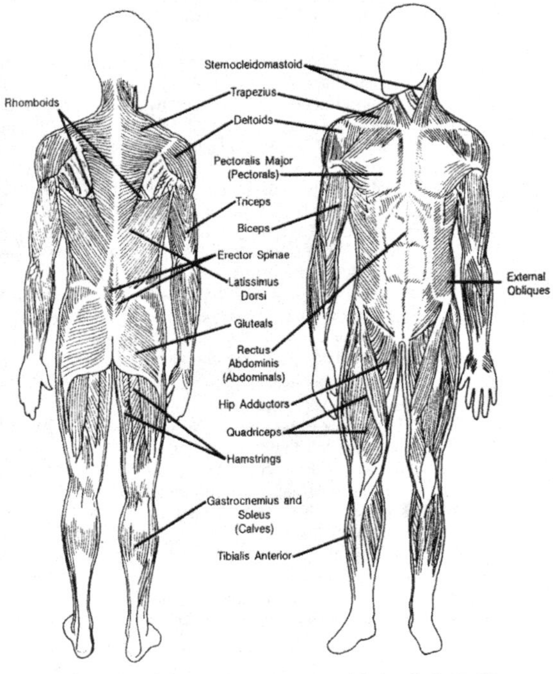

The iliopsoas muscle (a hip flexor) cannot be seen as it lies beneath other muscles. It attaches to the lumbar, the pelvis, the vertebrae and the femur.

Figure 3-4

BEGINNING EXERCISE PROGRAM

NAME OF EXERCISE	MAJOR MUSCLE GROUP(S) WORKED*
1. Leg press or squat	---Quadriceps, Gluteals
2. Leg curl	---Hamstrings
3. Heel raise	---Gastrocnemius
4. Bench press	---Pectorals, Triceps, Deltoids
5. Lat pull-down or pull-up	---Latissimus Dorsi, Biceps
6. Overhead press	---Deltoids, Triceps
7. Sit-up	---Rectus Abdominus, Iliopsoas, oblique muscles
8. Bent-leg dead-lift	---Erector Spinae, Quadriceps, Gluteals

Figure 3-5

The beginning weight-training program shown at Figure 3-5 will work most of the important, major muscle groups. It is a good program for beginners and for those whose time is limited. The exercises should be done in the order presented.

The weight-training program shown at Figure 3-6 is a more comprehensive program that works the major muscle groups even more thoroughly. It has some duplication with respect to the muscles that are worked. For example, the quadriceps are worked by the leg press/squat and leg extensions, and the biceps are worked by the seated row, lat pull-down, and biceps curl. Thus, for the beginner, this program may overwork some muscle groups. However, for the more advanced lifter, it will make the muscles work in different ways and from different angles thereby providing a better over-all development of muscle strength. This program also includes exercises to strengthen the neck muscles.

When doing one set of each exercise to muscle failure, the average soldier should be able to complete this routine and do a warm-up and cool-down within the regular PT time.

MORE ADVANCED EXERCISE PROGRAM

NAME OF EXERCISE	MAJOR MUSCLE GROUP(S) WORKED
1. Leg press or squat	---Quadriceps, Gluteals
2. Leg raises	---Iliopsoas (hip flexors)
3. Leg extension	---Quadriceps
4. Leg curl	---Hamstrings
5. Heel raise	---Gastrocnemius, Soleus
6. Bench press	---Pectorals, Triceps, Deltoids
7. Seated row	---Rhomboids, Latissimus dorsi, Biceps
8. Overhead press	---Deltoids, Triceps
9. Lat pull-down or pull-up	---Latissimus dorsi, Biceps
10. Shoulder shrug	---Upper trapezius
11. Triceps extension	---Triceps
12. Biceps curl	---Biceps
13. Sit-up	---Rectus abdominus, iliopsoas
14. Bent-leg dead lift	---Erector spinae, Quadriceps, Gluteals
15. Neck flexion	---Sternocleidomastoid
16. Neck extension	---Upper trapezius

Figure 3-6

Key Points to Emphasize

Some key points to emphasize when doing resistance training tire as follows

- Train with a partner if possible, This helps to increase motivation, the intensity of the workout, and safety,
- Always breathe when lifting. Exhale during the concentric (positive] phase of contraction, and inhale during the eccentric (negative) phase,
- Accelerate the weight through the concentric phase of contraction, and return the weight to the starting position in a controlled manner during the eccentric phase,
- Exercise the large muscle groups first, then the smaller ones.
- Perform all exercises through their full range of motion. Begin from a fully extended, relaxed position (prestretched), and end the concentric phase in a fully contracted position,
- Always use strict form. Do not twist, lurch, lunge, or arch the body, This can cause serious injury. These motions also detract from the effectiveness of the exercise because they take much of the stress off the targeted muscle groups and place it on other muscles.
- Rest from 30 to 180 seconds between different exercises and sets of a given exercise.
- Allow at least 48 hours of recovery between workouts, but not more than 96 hours, to let the body recover and help prevent over training and injury.
- Progress slowly, Never increase the resistance used by more than 10 percent at a time.
- Alternate pulling and pushing exercises. For example, follow triceps extensions with biceps curls.
- Ensure that every training program is balanced. Train the whole body, not just specific areas. Concentrating on weak areas is all right, but the rest of the body must also be trained.

Exercise Programs

When developing strength programs for units, there are limits to the type of training that can be done. The availability of facilities is always a major concern. Although many installations have excellent strength-training facilities, it is unreasonable to expect that all units can use them on a regular basis. However, the development of strength does not require expensive equipment. All that is required is for the soldier, three times a week, to progressively overload his muscles.

TRAINING WITHOUT SPECIAL EQUIPMENT

Muscles do not care what is supplying the resistance. Any regular resistance exercise that makes the muscle work harder than it is used to causes it to adapt and become stronger. Whether the training uses expensive machines, sandbags, or partners, the result is largely the same.

Sandbags are convenient for training large numbers of soldiers, as they are available in all military units. The weight of the bags can be varied depending on the amount of fill. Sandbag exercises are very effective in strength-training circuits. Logs, ammo boxes, dummy rounds, or other equipment that is unique to a unit can also be used to provide resistance for strength training. Using a soldier's own body weight as the resistive force is another excellent alternative method of strength training. Pull-ups, push-ups, dips, sit-ups, and single-leg squats are examples of exercises which use a person's body weight. They can improve an untrained soldier's level of strength.

Partner-resisted exercises (PREs) are another good way to develop muscular strength without equipment, especially when training large numbers of soldiers at one time. As with all training, safety is a critical factor. Soldiers should warm up, cool down, and follow the principles of exercise previously outlined.

PARTNER-RESISTED EXERCISE

In partner-resisted exercises (PREs) a person exercises against a partner's opposing resistance. The longer the partners work together, the more effective they should become in providing the proper resistance for each exercise. They must communicate with each other to ensure that neither too much nor too little resistance is applied. The resister must apply enough resistance to bring the exerciser to muscle failure in 8 to 12 repetitions. More resistance usual] y can and should be applied during the eccentric (negative) phase of contraction (in other words, the second half of each repetition as the exerciser returns to the starting position). The speed of movement for PREs should always be slow and controlled. As a general rule, the negative part of each exercise should take at least as long to complete as the positive part. Proper exercise form and regularity in performance are key ingredients when using PREs for improving strength.

Following are descriptions and illustrations of several PREs. They should be done in the order given to ensure that the exercising soldier is working his muscle groups from the largest to the smallest. More than one exercise per muscle group may be used. The PT leader can select exercises which meet the unit's specific goals while considering individual limitations:

A 36-to 48-inch stick or bar one inch in diameter may be used for some of the exercises. This gives the resister a better grip and/or leverage and also provides a feel similar to that of free weights and exercise machines.

SPLIT-SQUAT

This exercise is for beginning trainees' quadriceps and gluteal muscles.

Exerciser

Position: Stand erect with both feet pointed straight ahead, the left foot placed in a forward position and the right foot placed about 2.5 feet behind the left foot.

Action: Keeping the back straight and the head up, bend both legs at the same time, and lower yourself slowly until the right knee barely touches the floor. Return to the starting position. This is one repetition. After 8 to 12 repetitions to muscle failure, repeat the action with the opposite leg forward.

Resister

Position: Stand directly behind the exerciser with the fleshy portion of your forearms resting squarely on the exerciser's shoulders. You may clasp your hands to gain extra leverage as long as you do not squeeze the exerciser's neck. Be sure to place the same foot forward as the exerciser.

Action: As the exerciser lowers himself, apply a steady, forceful pressure downward against his shoulders. A slightly lesser pressure should be applied as the exerciser returns to the starting position.

SINGLE-LEG SQUAT

This exercise is for advanced trainees' quadriceps and gluteal muscles.

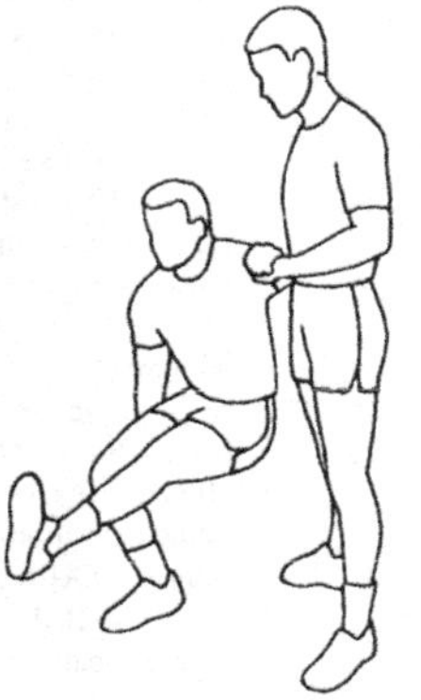

Exerciser

Position: Face your partner and grasp his wrists. Extend your right leg in front; keep it straight but do not let it contact your partner.

Action: Lower yourself in a controlled manner. Next, return to the upright position. After 8-12 repetitions to muscle failure, repeat this exercise with the other leg.

Resister

Position: Face the exerciser with your arms extended obliquely forward.

Action: Provide stability to the exerciser along with resistance or assistance as needed. When the exerciser can do more than 12 repetitions, apply an appropriate resistance that results in muscle failure in 8-12 repetitions.

LEG EXTENSION

This exercise is for the quadriceps muscles.

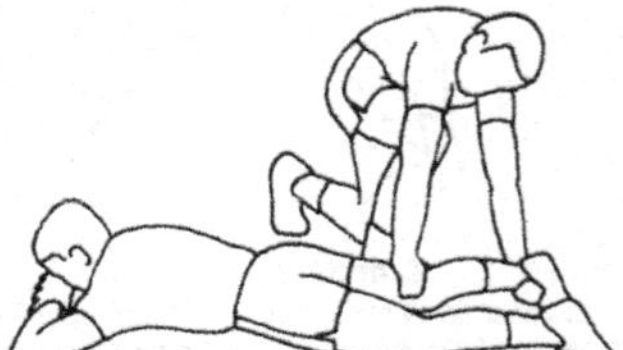

Exerciser

Position: Lie face down with one leg straight and the other flexed at the knee. Move your heel as close to your buttocks as possible.

Action: Extend your knee against the partner's resistance. Next, resist as your partner returns you to the starting position. Do 8 to 12 repetitions to muscle failure. Repeat this exercise with the other leg.

Resister

Position: Support the leg being exercised by placing your foot under the exerciser's thigh just above his knee.

Action: Resist while exerciser extends his leg. Next, apply upward pressure to return the exerciser to the starting position.

LEG CURL

This exercise is for the hamstring muscles.

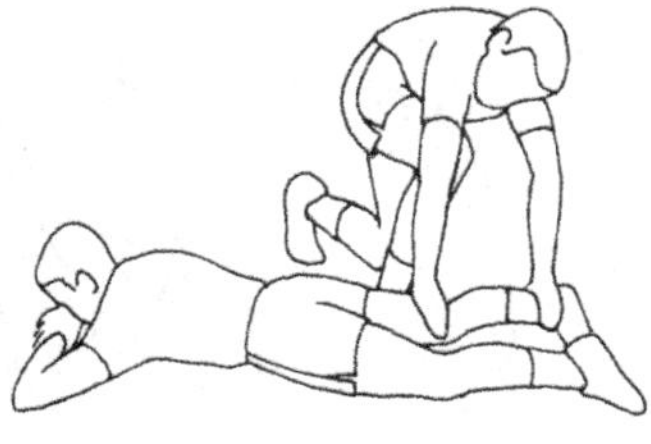

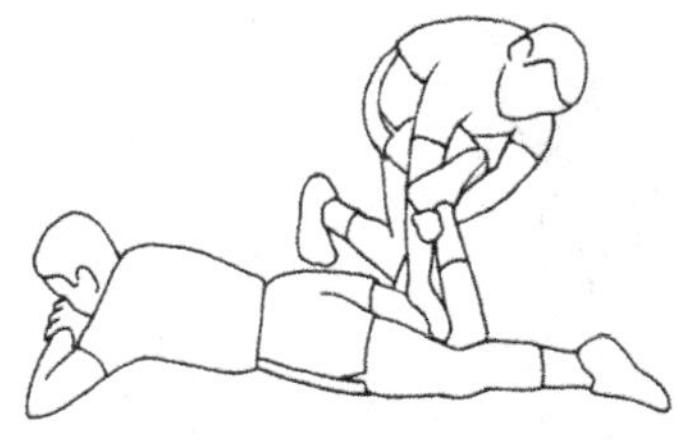

Exerciser

Position: Lie face down with your legs extended.
Action: Flex one leg against your partner's resistance until your heel is as close to your buttocks as possible. Next, resist your partner's efforts as he returns you to the starting position. Do 8 to 12 repetitions to muscle failure. Repeat this exercise with the other leg.

Resister

Position: Support the exerciser's leg as in the Leg Extension exercise.
Action: Resist the exerciser's movement with your hand(s) placed on his heel. Next, apply downward pressure to return the exerciser to the starting position.

HEEL RAISE (BENT OVER)

This exercise is for the gastrocnemius and soleus muscles.

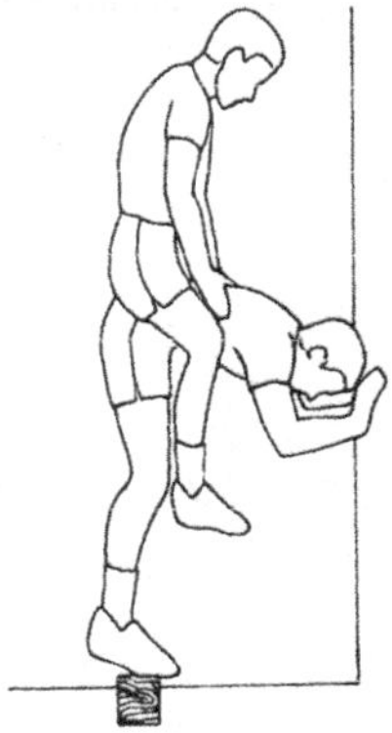

Exerciser

Position: Form a 90-degree angle between your upper body and legs by bending over at the hips. Use an additional partner or a fixed object for support.
Action: Keep your legs straight and rise up on the balls of your feet. Do 8 to 12 repetitions to muscle failure. If possible, perform the exercise by placing the balls of your feet firmly on a 4" x 4" board or the edge of a curb. Be sure to lower and raise your heels as far as possible.

Resister

Position: Sit on the upper part of the exerciser's buttocks; DO NOT SIT ON THE EXERCISER'S LOW BACK. (Properly positioning your body places less pressure on the exerciser's back and helps him better work his gastrocnemius and soleus muscles.)
Action: Provide resistance to the exerciser with your body weight.

TOE RAISE

This exercise is for the tibialis anterior muscle.

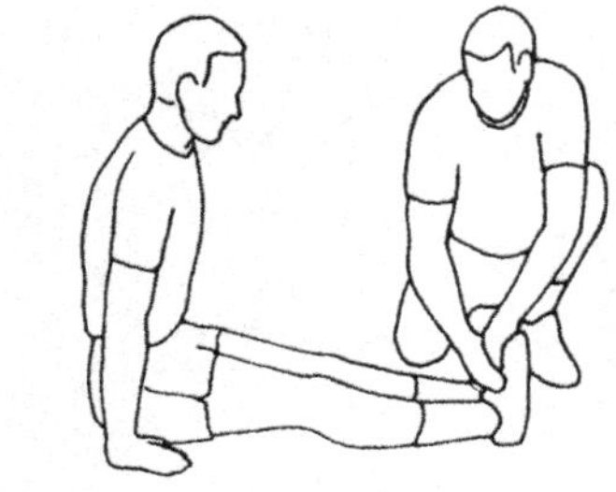

Exerciser

Position: Sit on the floor with your legs together, knees straight, and feet fully extended.

Action: Against the resister's efforts, move your toes toward the knees; then have the resister pull your toes back to the starting position while you resist. Do 8 to 12 repetitions to muscle failure.

Resister

Position: Place your hand(s) on the exerciser's shoelaces near the toes. Press your palms against the exerciser's insteps to resist his foot and ankle movements.

Action: Resist the exerciser's effort to pull his toes toward his knees. Next, pull the exerciser's toes back to the starting position against his resistance.

PUSH-UP

This exercise is for the pectoral and triceps muscles.

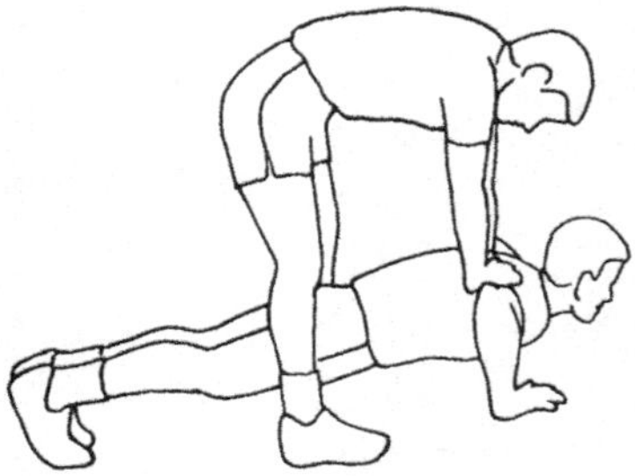

Exerciser

Position: Assume a front-leaning-rest position.

Action: Perform a push-up against your partner's resistance. Do 8 to 12 repetitions to muscle failure.

Resister

Position: Straddle the exerciser's hips. Place your hands on top of his shoulders. Be careful to place your left hand on the upper left part and your right hand on the upper right part of his shoulder.

Action: Apply pressure against the exerciser's push-up movements. As stated earlier, slightly more resistance should be applied during the eccentric phase of contraction (in this case, as the exerciser moves closer to the floor)

SEATED ROW

This exercise is for the biceps, latissimus dorsi, and rhomboid muscles.

Exerciser

Position: Sit facing the resister with your back straight. Overlap your legs with the resister's, being sure to place your legs on top. Establish a good grip by interlocking your hands with the resister's or by firmly grasping his wrists. The exerciser's palms should be facing downward.

Action: Pull the resister toward you with a rowing motion while keeping your elbows elevated to shoulder height. Be sure to keep your back straight, and move only the arms.Next, slowly return to the starting position as the resister pulls your arms forward. Do 8 to 12 repetitions to muscle failure.

Resister

Position: Face the exerciser and sit with your back straight. Place your legs under the exerciser's legs; establish a good grip by interlocking hands with the resister or by firmly grasping his wrists.

Action: As the exerciser pulls, resist his pulling motion. Next, slowly pull the exerciser back to the starting position by pulling with the muscles of the lower back.

OVERHEAD PRESS

This exercise is for the deltoid and triceps muscles.

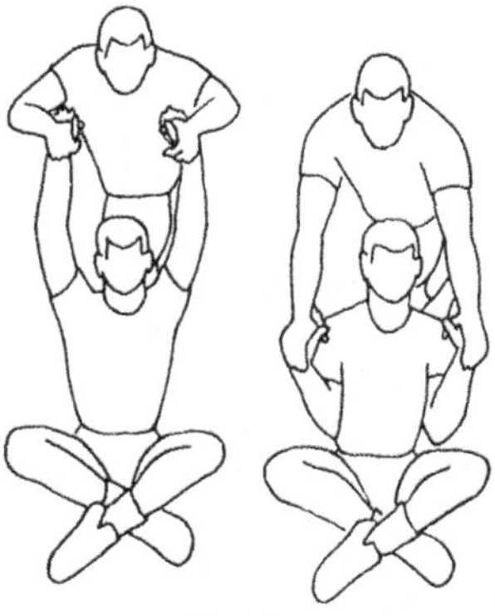

Exerciser

Position: Sit with your legs crossed and your back straight. Raise your hands to shoulder height with your palms flat and facing upward.

Action: Move your arms slowly upward to full extension against your partner's resistance. Next, slowly return to the starting positicn as the resister applys downward pressure. Do 8 to 12 repetitions to muscle failure.

Resister

Position: Stand behind the exerciser; interlock your thumbs with the exerciser's, and place your hands with the palms down on his hands. Support the exerciser's back with the side of your lower leg.

Action: Resist the exerciser's upward movement; then push his arms back to the starting position. A bar or stick may be used for a better grip and improved leverage.

PULL-DOWN

This exercise is for the latissimus dorsi muscles.

Exerciser

Position: Sit with your legs crossed and back straight. Raise and cross your arms behind your head with your elbows bent.
Action: Pull out and down with your elbows against the partner's resistance until your elbows touch your ribcage. Next, resist as your partner pulls your elbows back to the starting position. Do 8 to 12 repetitions to muscle failure.

Resister

Position: Stand behind the exerciser, and support his back with the side of your lower leg. Place your palms underneath the exerciser's elbows.
Action: Resist the exerciser's downward movements; then pull his elbows back to the up or starting position. VARIATION: A bar or stick may be used for a better grip and leverage and to exercise the biceps and forearm muscles.

SHRUG

This exercise is for the upper trapezius muscle.

Exerciser

Position: Sit with your legs crossed, back straight, and hands resting in your lap.
Action: Shrug your shoulders as high as possible against your partner's resistance, then resist your partner's pushing motion as you return to the starting position. Do 8 to 12 repetitions to muscle failure.

Resister

Positon: Stand behind the exerciser, and support his back with the side of your lower leg. Place your hands on each of the exerciser's shoulders.
Action: Apply pressure downward with your hands to resist the upward, shrugging movements of the exerciser and, during the second phase of the exercise, push downward as the exerciser resists your pushing movements.

TRICEPS EXTENSION

This exercise is for the triceps muscles.

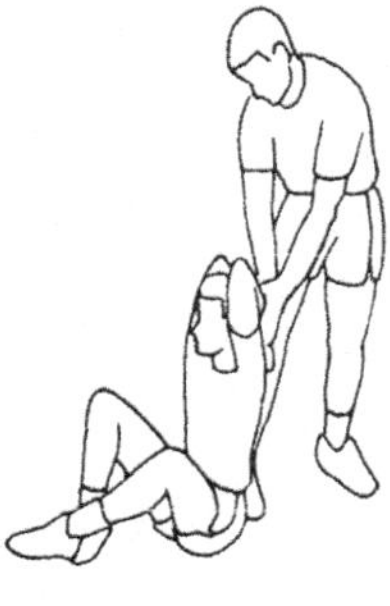

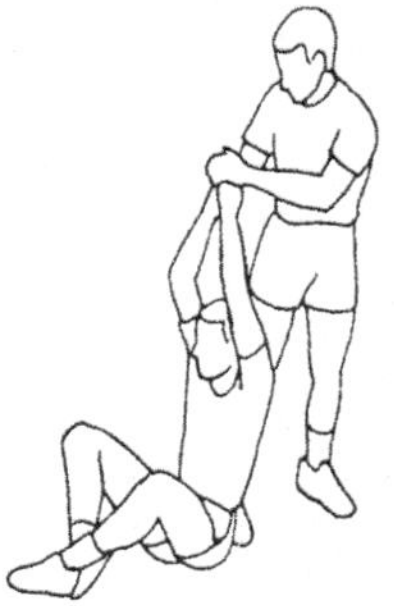

Exerciser

Position: Sit with your legs crossed and back straight. Clasp your hands and place them behind your head while bending your elbows.

Action: Extend your arms upward against the partner's resistance. Next, return to the starting position while resisting your partner's force. Always keep your elbows stationary and pointing straight ahead. Do 8 to 12 repetitions to muscle failure.

Resister

Position: Stand behind the exerciser and support his back with the side of your lower leg. Place your hands, palms down, over the exerciser's hands.

Action: Apply pressure to resist the upward movement of the exerciser, and then push his hands back to the starting position. A bar or a stick may be used for a better grip and/ or improved leverage. This exercise may also be done in the prone position with the resister applying a force against the exerciser's movements.

BICEPS CURL

This exercise is for the biceps muscles.

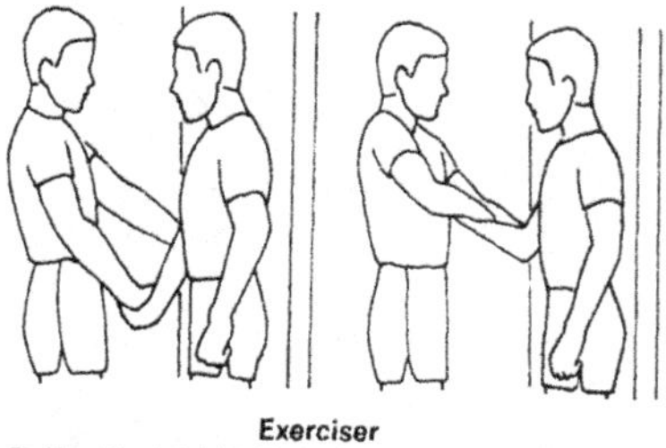

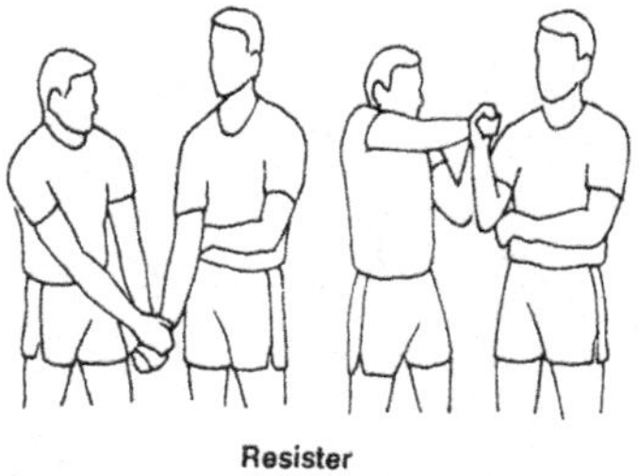

Exerciser

Position: Stand straight with your back supported. Hold the arm to be exercised close to your side.

Action: Bend the elbow, bringing your hand up to your shoulder against your partner's resistance. Return to the starting position by resisting the pushing efforts of your partner. Do 8 to 12 repetitions to muscle failure; repeat with the other arm.

Resister

Position: Face the exerciser with your feet staggered. Use one of your hands to grasp the exerciser's wrist; place the other hand behind his elbow to stabilize it during the exercise movement.

Action: Resist the exerciser's upward movement and provide a downward, pushing force during the lowering movement. A bar may also be used for a better grip and leverage.

VARIATION: A variation may be used if the resister is unable to provide enough resistance to the exerciser with the first exercise. Using this variation, the exerciser places the back of his hand *on the non-exercising arm behind the elbow of his exercising arm* for support; the resister places both hands on the hand, wrist, or lower part of the exerciser's forearm to apply resistance to the exerciser's movements. The action is the same as before.

ABDOMINAL CURL

This exercise is for the rectus abdominus, iliopsoas, and external and internal oblique muscles.

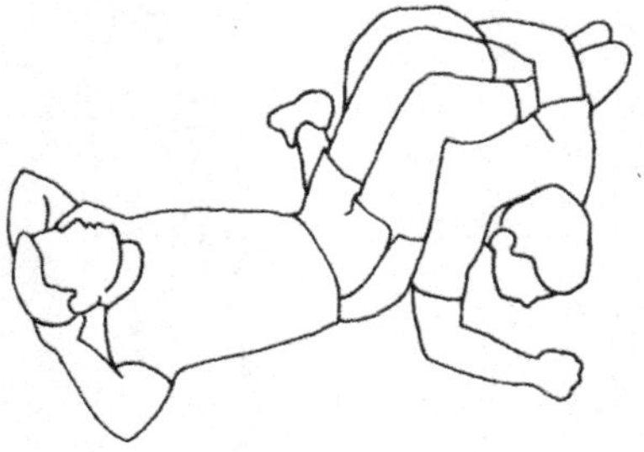

Exerciser

Position: Lie on your back with both legs bent at the knee to about a 90-degree angle. Place your bent legs over the resister's back. Interlace your fingers behind your neck.

Action: Do regular sit-ups, bringing both elbows to your knees. Do 20 to 50 repetitions to muscle failure.

NOTE: A variation to this exercise is the ROCKY SIT-UP where the exerciser moves the left elbow up to the right knee and then reverses the action, right elbow to left knee.

Resister

Position: Kneel with your inside elbow resting on the ground. With your outside arm, reach back and hold the exerciser's ankles.

Action: Provide a firm foundation upon which the exerciser can place his legs, and keep them tightly anchored during the exercise.

ABDOMINAL CRUNCH

This exercise is for the rectus abdominis and external and internal oblique muscles.

Exerciser

Position: Lie down with your arms crossed over your chest, the backs of your lower legs resting over your partner's back, and your upper leg placed at right angles to the floor.

Action: Curl your neck off the ground, and curl your upper body up toward your knees. (Progressively lift your shoulders, upper back, and finally, lower back off the ground.) Hold this position briefly while forcefully tensing your abdominal muscles. Return slowly to the starting position and repeat. Do 20 to 50 repetitions to muscle failure.

Resister

Position: Kneel with both forearms on the ground.

Action: Allow the exerciser to place the back of his lower legs on your back. DO NOT HOLD HIS LEGS DOWN. (This eliminates the iliopsoas muscle from the exercise and instead isolates the rectus abdominis and external and internal oblique muscles.)

TRAINING WITH EQUIPMENT

Units in garrison usually have access to weight rooms with basic equipment for resistance-training exercises. The exercises described here require free weights and supporting equipment. Although not shown below for the sake of simplicity, all exercises done with free weights require a partner, or spotter, to ensure proper form and the safety of the lifter.

Free-Weight Exercises

SQUAT

This exercise is for the quadriceps and gluteal muscles.

Position: Stand with the feet about shoulder width apart. Hold the weight on your shoulders.
Action: Bend the knees until the tops of your thighs are parallel to the ground. Keep your head and shoulders upright and back straight. In the lowest position, the top of your thighs should not go lower than parallel to the ground. Do 8 to 12 repetitions to muscle failure. A 2" x 4" block may be placed under the heels to increase stability.

HEEL RAISE

This exercise is for the gastrocnemius and soleus muscles.

Position: Place a bar on your shoulders behind your neck. Stand with the toes and the balls of the feet on a platform or a 4" x 4" board.

Action: Rise upward on the toes and balls of the feet to full extension, then slowly lower the heels as far as possible. Do not bend the knees or jerk the hips. Do 8 to 12 repetitions to muscle failure.

BENCH PRESS

This exercise is for the pectoralis major, triceps, and deltoid muscles.

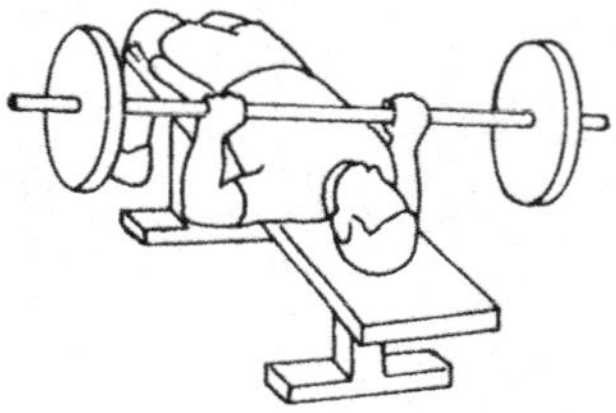

Position: Hold a weight with an overhand grip (palms facing away) slightly wider than shoulder width. Hold the bar directly above your chest at arm's length.

Action: Lower the bar to your chest, keeping the feet flat on the floor. Push the bar up to arm's length. The elbows should be kept wide and away from the body. Keep the buttocks in contact with the bench at all times. Do 8 to l2 repetitions to muscle failure.

BENT-OVER ROW

This exercise is for the latissimus dorsi and biceps muscles.

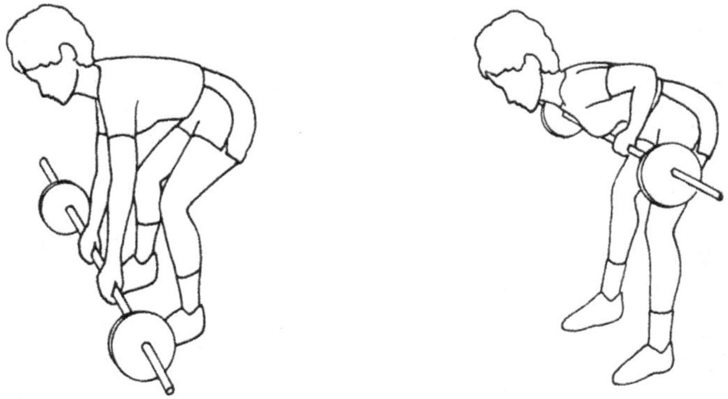

Position:. Lean forward at the hips with the back flat; let your arms hang straight down from the shoulders. Keep your knees slightly flexed.

Action: Use an overhand grip with the hands 12 to 24 inches apart. Bend the elbows, bringing the bar up in a straight motion up to the lower portion of the chest. Slowly lower the weight back to the starting position. Do 8 to 12 repetitions to muscle failure.

OVERHEAD PRESS

This exercise is for the deltoids and triceps muscles.

Position: With a barbell, use the overhand grip with the hands spaced slightly greater than shoulder width apart.

Action: Push the bar overhead, moving it upward in a straight line until the elbows are straight. Lower the bar until it touches the chest. Do not bounce the bar off the chest. Dumbbells may also be used. Do 8 to 12 repetitions to muscle failure.

SHRUG

This exercise is for the trapezius muscles.

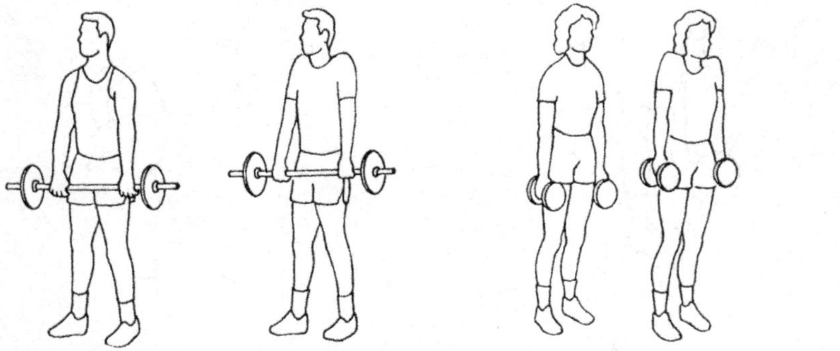

Position: Using a barbell, start with the bar at thigh-rest position. Use an overhand or reverse grip.

Action: Elevate the bar by contracting the trapezius and raising the shoulders upward toward the ears. In the top position, roll your shoulders backward. Then, slowly lower the shoulders until the bar returns to the starting, thigh-rest position. Keep the arms straight throughout the entire repetition. Dumbbells may also be used. Do 8 to 12 repetitions to muscle failure.

TRICEPS EXTENSION

This exercise is for the triceps muscles.

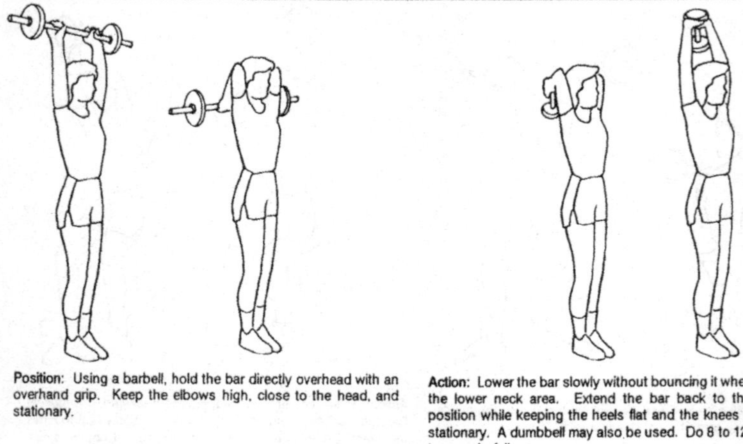

Position: Using a barbell, hold the bar directly overhead with an overhand grip. Keep the elbows high, close to the head, and stationary.

Action: Lower the bar slowly without bouncing it when it reaches the lower neck area. Extend the bar back to the overhead position while keeping the heels flat and the knees and elbows stationary. A dumbbell may also be used. Do 8 to 12 repetitions to muscle failure.

BICEPS CURL

This exercise is for the biceps muscles.

Position: Start with an underhand grip. Hold the bar at thigh-rest position.

Action: Keep the elbows stationary and close to your sides as you curl the bar to your chest. Do not use your legs or bend your back for assistance. A cambered (bent) bar or dumbbells may also be used. Do 8 to 12 repetitions to muscle failure.

WRIST CURL

This exercise is for the development of the forearm muscles.

Position: Holding your hand with the palms facing upward, grasp a barbell using only the fingers.

Action: Curl the fingers, then the wrist up as far as possible and then down, keeping the elbows stationary. For the best results, do not grip the barbell; keep it placed on the last few digits of the fingers. Do 8 to 12 repetitions to muscle failure.

BENT-LEG DEAD-LIFTS

This exercise is for the quadriceps, the erector spinae, the gluteals, and the trapezius muscles.

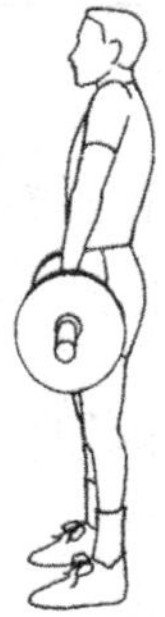

Position: Bend and grasp the bar with the hands shoulder width apart. The legs should be bent, the back flat but inclined forward at a 45 degree angle, the arms straight, and the head up.

Action: Keeping the head erect, gradually straighten the legs and the back together at the same time. Make sure that the back remains flat and the arms remain straight. When the entire body is straight, shrug the shoulders upward as high as possible. In a controlled manner, return to the starting position by first lowering the shoulders. Then, bend at the knees and at the waist simultaneously until the beginning position is attained. Keep the back flat, head up, and the arms straight at all times. Do 8 to 12 repetitions to muscle failure.

Exercises Performed with an Exercise Machine

If exercise machines are available, the exercises described below are also good for strength training. All movements, particularly during the eccentric (negative) phase of contraction, should be done in a deliberate, controlled manner.

LEG PRESS

This exercise is for the gluteal and quadriceps muscles.

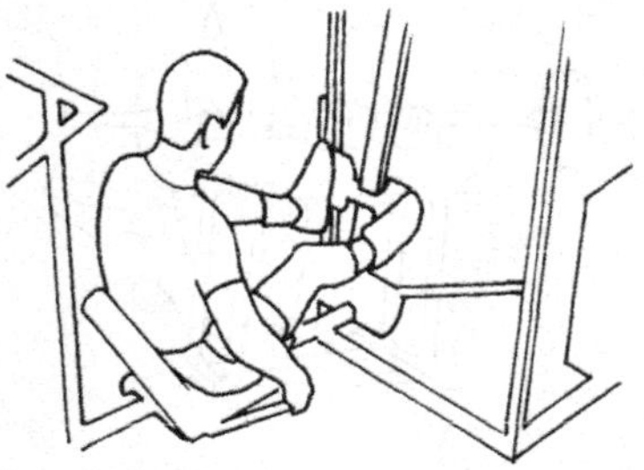

Position: Sit at the leg-press station with the legs bent no more than 90 degrees. Ensure that the balls of both feet are very securely placed on the pedals.

Action: Push the weight with the legs until your knees are straight but not locked. In a controlled manner, return to the starting position. This is one repetition. Do 8 to 12 repetitions to muscle failure.

LEG EXTENSION

This exercise is for the quadriceps muscles.

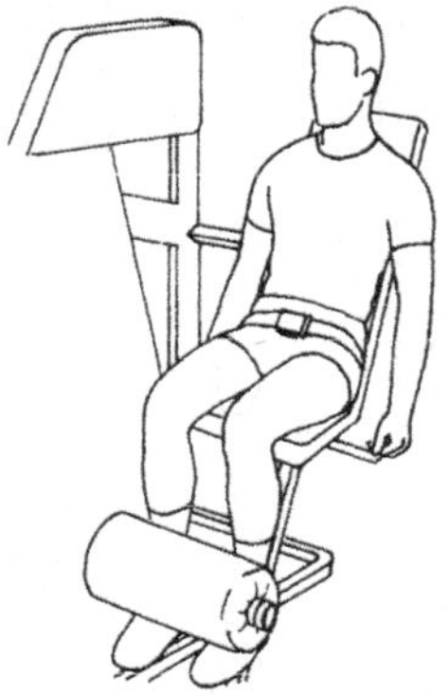

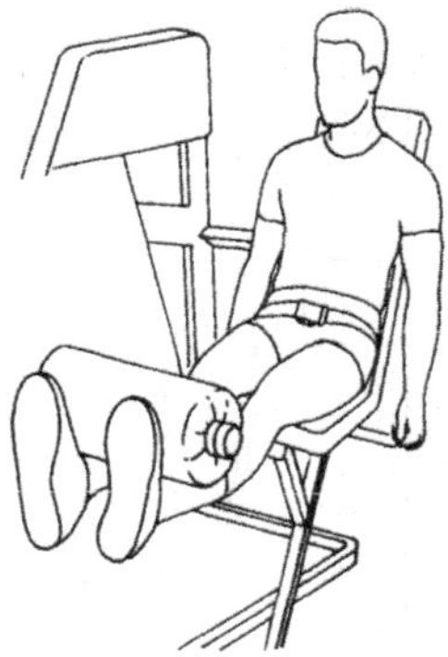

Position: Sit on a bench with your lower legs behind the padded lever. Hold on to the bench or provided handles with your hands to keep the upper body in the correct position.

Action: Straighten the legs as much as possible. In a controlled manner, return to the starting position. This is one repetition. Do 8 to 12 repetitions to muscle failure.

LEG CURL

This exercise is for the hamstring muscles.

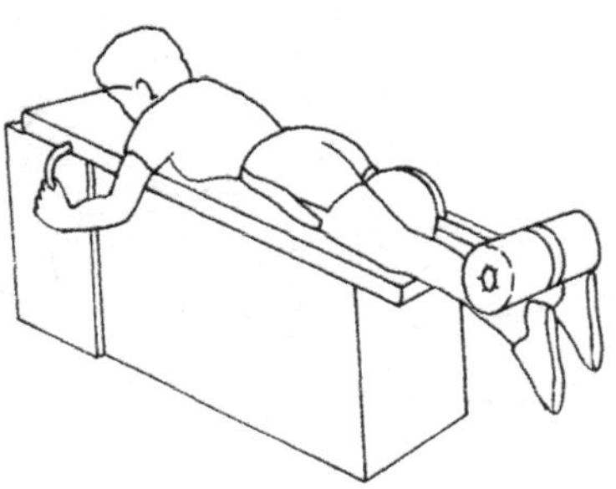

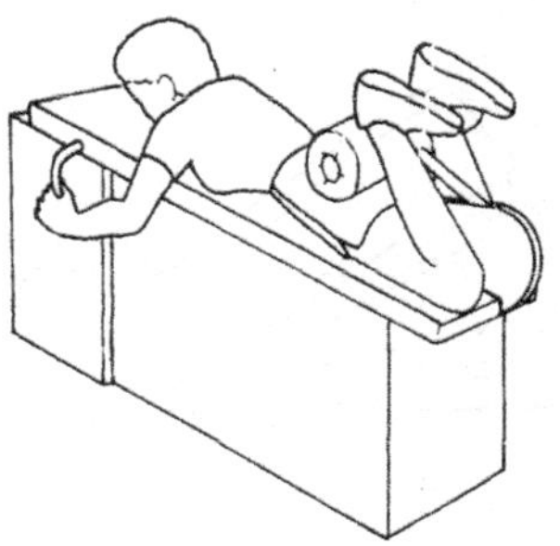

Position: Lie on your stomach with the legs straight and the ankles under the padded lever. Maintain correct upper body position by loosely grasping the sides of the bench or provided handles.

Action: Bend your legs at the knee until the lower legs pass well beyond the perpendicular position and the heels are as close to your buttocks as possible. Return to the starting position. Do 8 to 12 repetitions to muscle failure.

HEEL RAISE

This exercise is for the gastrocnemius and soleus muscles.

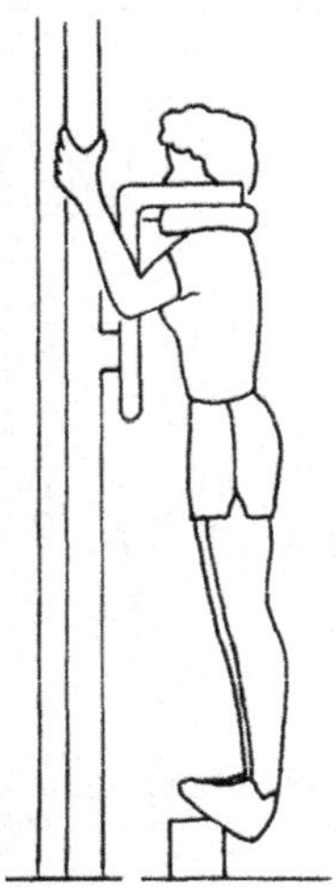

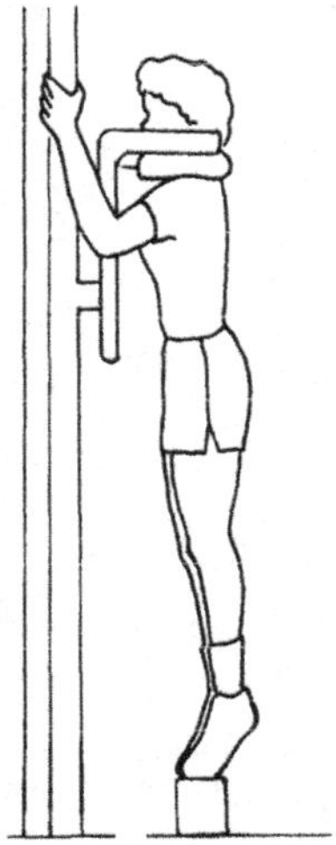

Position: Stand with a weight on your shoulders and the balls of your feet placed firmly on a 4-inch raised surface.

Action: Raise your heels off the floor as far as possible while maintaining your balance. Then, lower them as far as possible. This is one repetition. Do 8 to 12 repetitions to muscle failure.

TOE RAISE

This exercise is for the tibialis anterior muscle.

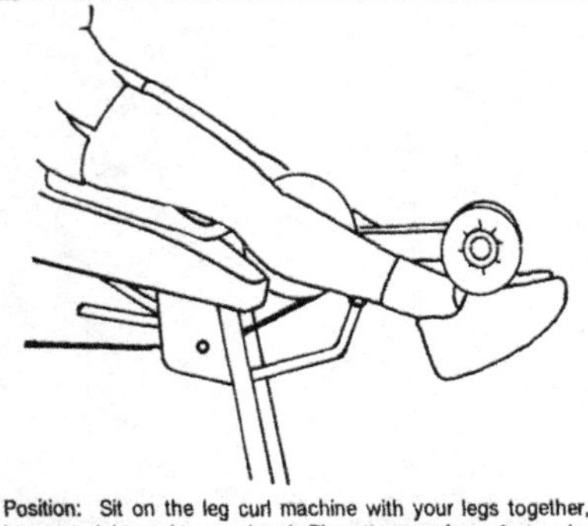

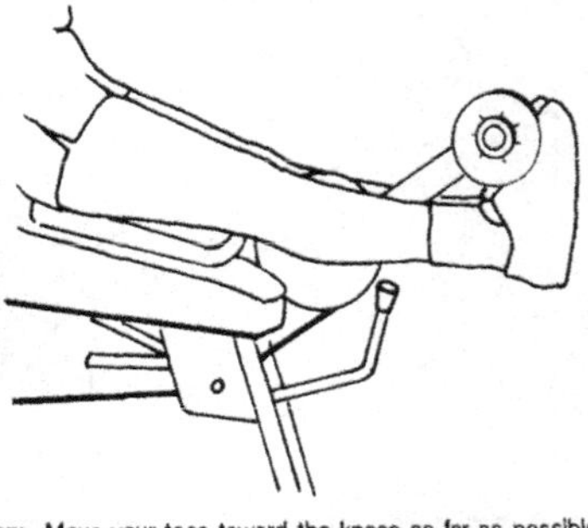

Position: Sit on the leg curl machine with your legs together, knees straight, and toes pointed. Place the top of your feet under the roller pad.

Action: Move your toes toward the knees as far as possible. Then lower the weight to the starting position in a controlled manner. Do 8 to 12 repetitions to muscle failure.

BENCH PRESS

This exercise is for the pectoralis major, triceps, and deltoid muscles.

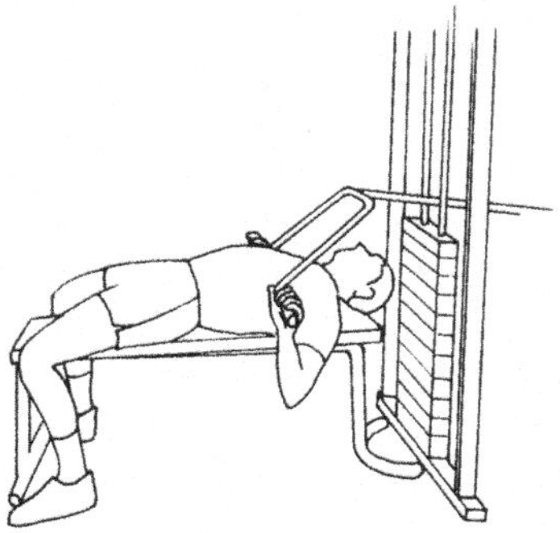

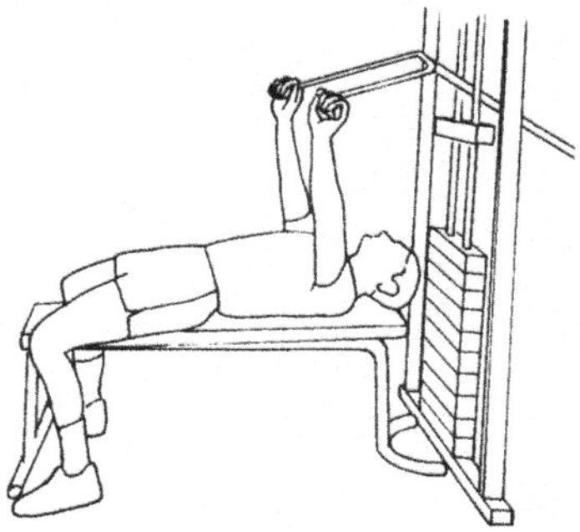

Position: Lie on your back with your hands placed about shoulder width apart on the bar. Generally, the bar or handles should be located at the lower half of the chest.

Action: Push the bar up until your arms are straight. Then, lower the bar to the starting position. This is one repetition. Do 8 to 12 repetitions to muscle failure.

SEATED ROW

This exercise is for the latissimus dorsi and biceps muscles.

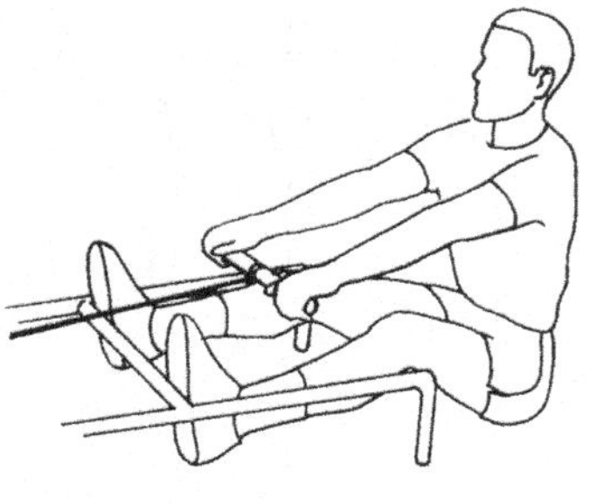

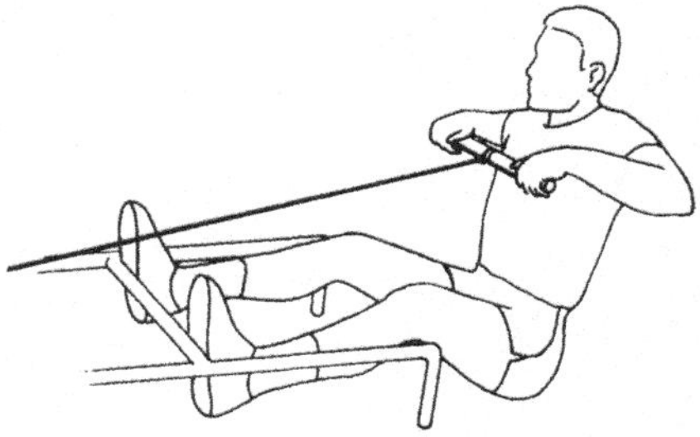

Position: Sit and assume the straight arm position shown above. Use the overhand grip with your hands spaced 6 to 8 inches apart.

Action: Pull the bar to the lower part of your chest, while keeping your elbows elevated to shoulder height, then slowly extend the arms and lower the weight to the beginning position. Be sure to keep the back straight, and move only the arms. Do 8 to 12 repetitions to muscle failure.

LAT PULL-DOWN

This exercise is for the latissimus dorsi and biceps muscles. (Pull-ups or chin-ups may be substituted for this exercise.)

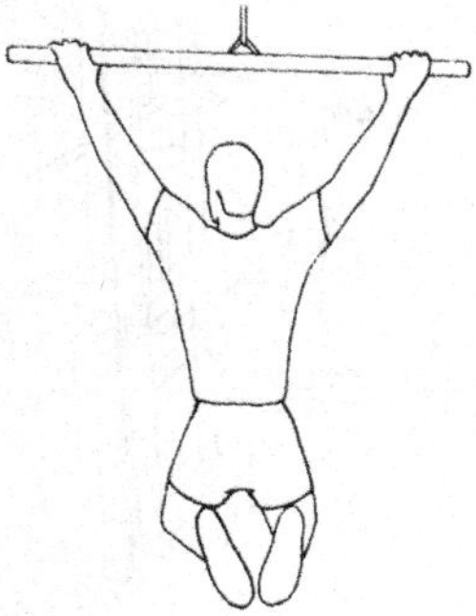

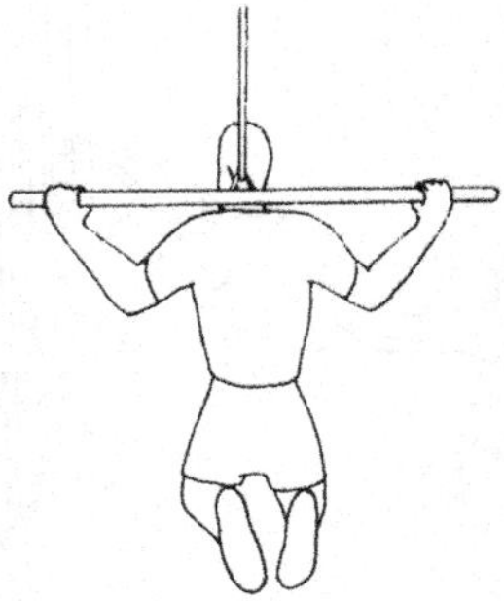

Position: Sit or kneel and grasp the bar with your palms facing away from the body.

Action: Pull the bar down until it touches the back of your neck; return the bar in a controlled manner to that starting position. This is one repetition. Do 8 to 12 repetitions to muscle failure.

SHRUG

This exercise is for the trapezius muscles of the upper back.

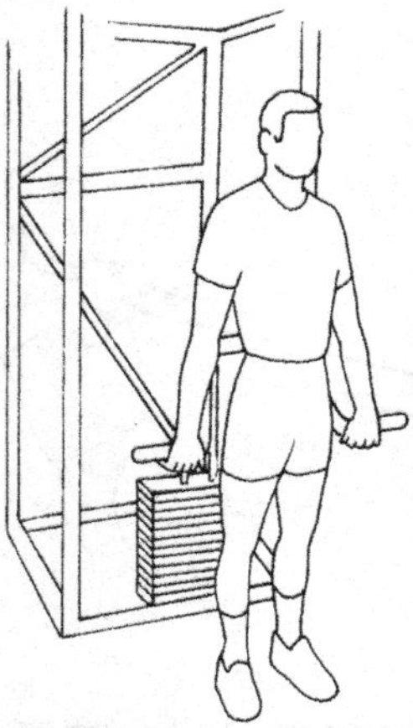

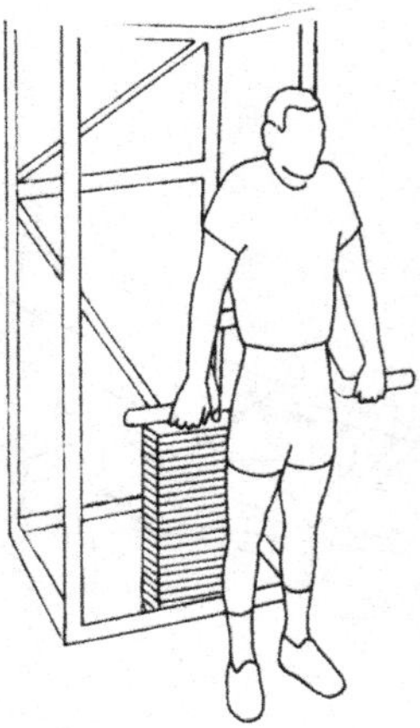

Position: Stand with the feet shoulder width apart. Hold a weight in your hands with the arms locked in a straight position.

Action: Pull the shoulders up toward your ears as far as possible and then backward. Always keep your arms completely straight. Next, lower your shoulders to the starting position. This is one repetition. Do 8 to 12 repetitions to muscle failure.

PARALLEL BAR DIP

This exercise is for the pectoralis major and triceps muscles.

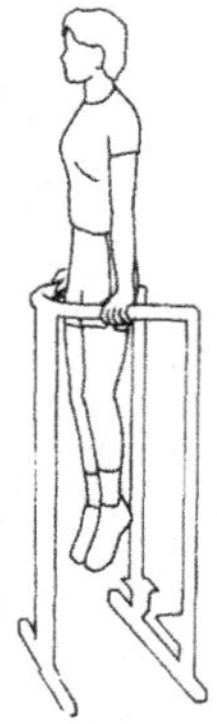

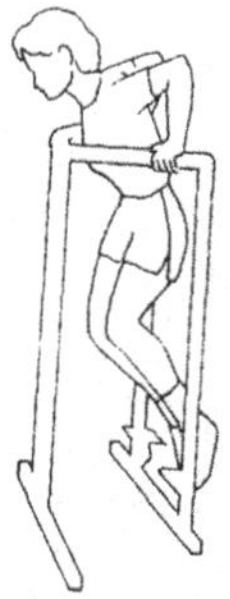

Position: Keep your feet off the floor and support the body's weight on straight arms.

Action: Bend the arms and lower your body until the upper arms are at least parallel to the floor. If necessary, bend your legs at the knees to keep the feet from touching the floor. Straighten your arms to return to the starting position. This is one repetition. Do 8 to 12 repetitions to muscle failure. A weight belt may be worn if additional resistance is needed.

CHIN-UP

This exercise is for the latissimus dorsi and biceps muscles. (Lat pull-downs or pull-ups may be substituted for this exercise.)

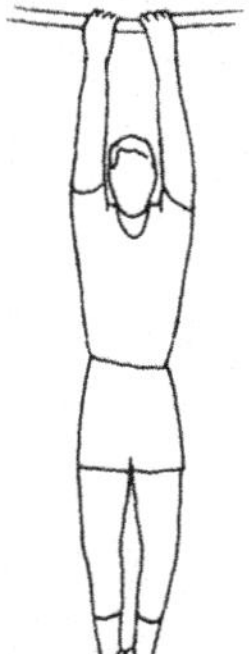

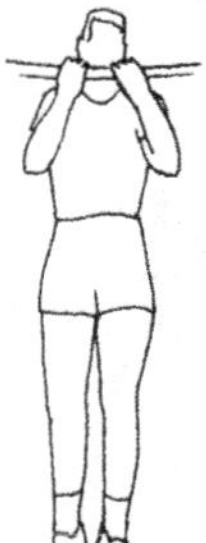

Position: From a standing position, grasp the bar with your palms facing the body.

Action: Bending both arms, pull your body up until your chin clears the bar. Return to the starting position in a controlled manner. If necessary, bend your knees to keep the feet from touching the floor. Do 8 to 12 repetitions to muscle failure. A weight belt may be worn if additional resistance is needed.

TRICEPS EXTENSION

This exercise is for the triceps muscles.

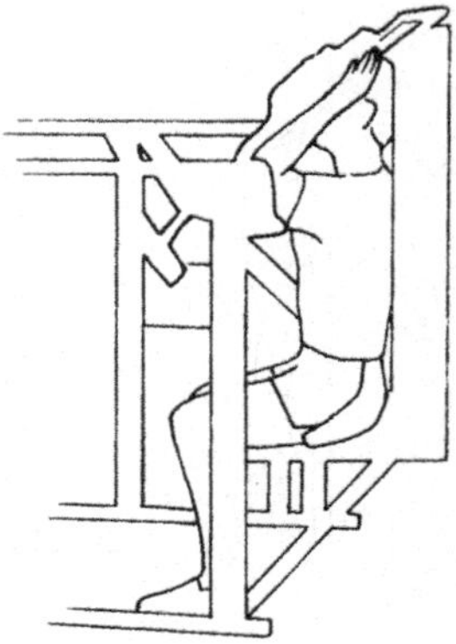

Position: Make seat adjustments to sit with the elbows in line with the axes of the cams and the elbows slightly higher than the shoulders. Place the sides of the hands on the pads with the hands open.

Action: Straighten the arms against the resistance. After doing this, bend the elbows, and return to the starting position in a controlled manner. Do 8 to 12 repetitions to muscle failure.

BICEPS CURL

This exercise is for the biceps muscles.

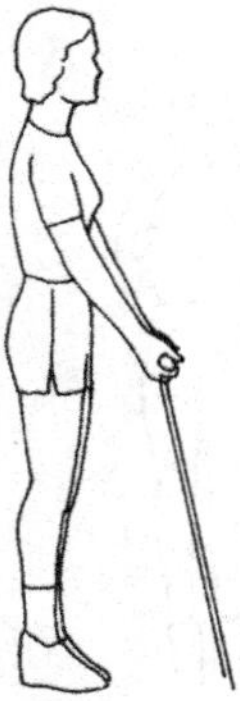

Position: Stand with the bar in front of your body, arms straight and elbows at the sides. Your hands should be spaced about shoulder width apart and the palms should face away from the body.

Action: Without moving your elbows, bend the arms, bringing the bar to shoulder level. In a controlled manner, lower the weight to the starting position. This is one repetition. Do 8 to 12 repetitions to muscle failure.

The following exercises can be performed to condition the muscles of the mid-section (erector spinae, rectus abdominus and external and internal obliques). As the soldier becomes more conditioned on these exercises, resistance can be added.

BACK EXTENSION

This exercise is for the erector spinae muscle group.

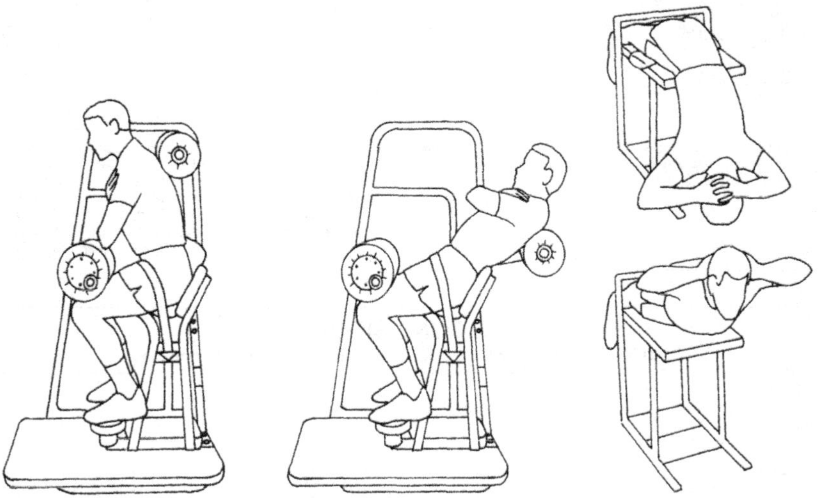

Position: Sit in the machine with your back underneath the highest roller pad. Stabilize your lower body by moving your thighs under the lower roller pads. Place the feet firmly on the platform and fasten the seat belt. Interlace your fingers across your waist, or fold your arms across your chest.

Action: Move the torso backward smoothly until the upper body forms a straight line with the lower body. Do not arch the back excessively by moving past this point. Return to the starting position in a controlled manner. Do 8 to 12 repetitions to muscle failure.

Alternative: *If a low-back machine is not available, the following exercise may be performed.*

Position: While face-down, anchor your feet securely and position the supporting pad under the upper part of the front thighs. Position your upper body as close to vertical as possible. The hands may be placed behind the head with fingers interlocked; or, the arms may be folded across the chest, provided they do not restrict the downward range of motion.

Action: Straighten the back and raise the upper body until it forms a straight line with the legs. Do not allow your upper body to come any higher than parallel to the floor. Lower your upper body to the starting position in a controlled manner. Do 8 to 12 repetitions to muscle failure.

SIT-UP

This exercise is for the rectus abdominis and iliopsoas (hip flexor) muscles.

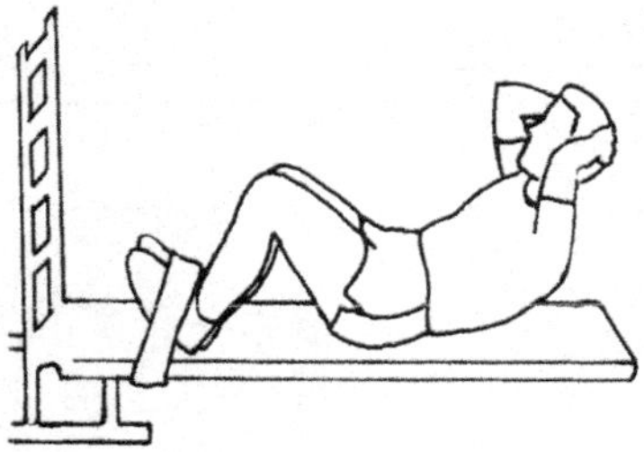

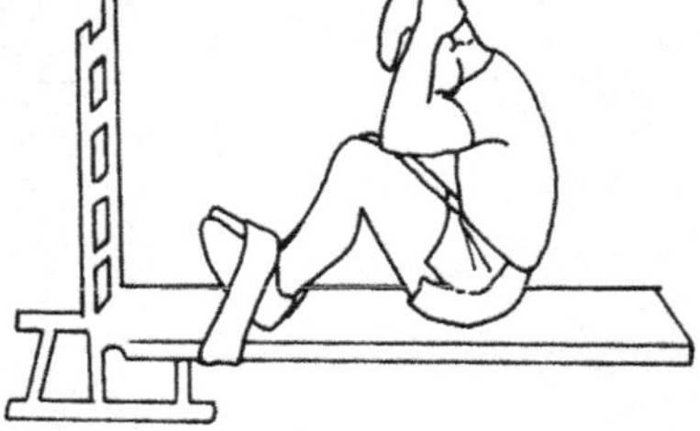

Position: Lie on your back with your knees bent at approximately a 90 degree angle and feet anchored. Place your hands behind your head.

Action: Sit up until your trunk is in a vertical position relative to the floor while keeping the knees bent. Lower yourself in a controlled manner to the starting position. The number of repetitions you should do depends on the maximum number of sit-ups you perform in two minutes. Do three sets of 50 percent of your maximum number. For example, if you can do 60 sit-ups in two minutes, do three sets of 30 or more repetitions per set.

INCLINE SIT-UP

This exercise is for the rectus abdominis and iliopsoas muscles.

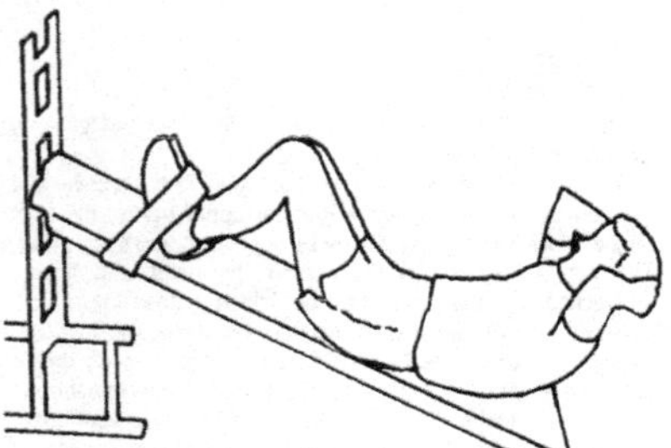

Position: Lie on an incline board with your knees bent at approximately a 90 degree angle and your feet anchored. The steeper the incline of the board, the more difficult the sit-up will be. Interlace the fingers behind your head.

Action: Curl your torso up as far as comfortably possible. Return to the starting position. This is one repetition. Do 20 to 50 repetitions to muscle failure.

ABDOMINAL CRUNCH

This exercise is for the rectus abdominis muscle.

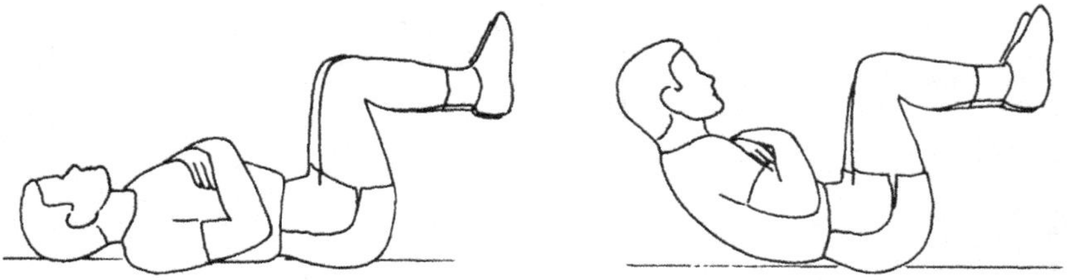

Position: Lie on your back with both legs bent at the knees and the upper legs at right angles to the floor. Your arms should be crossed at chest level with the palms of the hands on their opposite shoulders. Your ankles may be crossed but, in all cases, the feet should not touch the floor.

Action: Roll up your upper body by first lifting your head and tucking the chin. Next, curl your spine by rolling the upper back and then the lower back off the floor. Pause briefly in the up position while tensing the abdominal muscles. Return in a slow, controlled manner to the starting position by "unrolling" the upper body. Do 20 to 50 repetitions to muscle failure.

Exercise Chart

The chart labeled Figure 3-5 will help the soldier select appropriate exercises for use in developing a good muscular endurance and strength workout. For example, if the soldier wants to develop his upper leg muscles, he has several options. He may choose from the following: 1) PREs, concentrating on the split- or single-leg squat; 2) exercises with equipment, doing free weight squats; or, 3) exercises with a machine, doing leg presses, leg curls, and leg extensions.

EXERCISE CHART FOR MUSCULAR STRENGTH AND ENDURANCE

	EXERCISES	LOWER LEGS	UPPER LEGS	WAIST	CHEST	UPPER ARMS	LOWER ARMS	SHOULDERS	BACK
Partner-Resisted Exercises	Split-Squat		x						
	Single-Leg Squat		x						
	Leg Extension		x						
	Leg Curl		x						
	Heel Raise	x							
	Toe Raise	x							
	Push-Up				x	x			
	Seated Row					x			x
	Overhead Press					x		x	
	Pull-Down					x			x
	Shrug							x	
	Triceps Extension					x			
	Biceps Curl					x			
	Abdominal Twist			x					
	Abdominal Curl			x					
	Abdominal Crunch			x					
Exercises with Equipment (Barbell/Dumbbell)	Squat		x						
	Heel Raise	x							
	Bench Press				x	x			
	Bent-Over Row					x			x
	Overhead Press					x		x	
	Shrug							x	
	Triceps Extension					x			
	Biceps Curl					x			
	Wrist Curl						x		
	Bent-Leg Dead Lift		x					x	x
Exercises with an Exercise Machine	Leg Press		x						
	Leg Extension		x						
	Leg Curl		x						
	Heel Raise	x							
	Toe Raise	x							
	Bench Press				x	x			
	Seated Row					x			x
	Lat Pull-Down					x			x
	Shrug							x	
	Parallel Bar Dip				x	x			
	Chin-up					x			x
	Triceps Extension					x			
	Biceps Curl					x			
	Back Extension								x
	Sit-Up			x					
	Incline Sit-Up			x					
	Abdominal Twist			x					
	Abdominal Crunch			x					

Figure 3-5

CHAPTER 4

Flexibility

Flexibility refers to the range of movement of a joint.

The four categories of stretching techniques are static,passive, proprioceptive neuromuscular facilitation (PNF), and ballistic.

Flexibility is a component of physical fitness. Developing and maintaining it are important parts of a fitness program. Good flexibility can help a soldier accomplish such physical tasks as lifting, loading, climbing, parachuting, running, and rappelling with greater efficiency and less risk of injury.

Flexibility is the range of movement of a joint or series of joints and their associated muscles. It involves the ability to move a part of the body through the full range of motion allowed by normal, disease-free joints.

No one test can measure total-body flexibility. However, field tests can be used to assess flexibility in the hamstring and low-back areas. These areas are commonly susceptible to injury due, in part, to loss of flexibility. A simple toe-touch test can be used. Soldiers should stand with their legs straight and feet together and bend forward slowly at the waist. A soldier who cannot touch his toes without bouncing or bobbing needs work to improve his flexibility in the muscle groups stretched by this test. The unit's Master Fitness Trainer can help him design a stretching program to improve his flexibility.

Stretching during the warm-up and cool-down helps soldiers maintain overall flexibility. Stretching should not be painful, but it should cause some discomfort because the muscles are being stretched beyond their normal length. Because people differ somewhat anatomically, comparing one person's flexibility with another's should not be done. People with poor flexibility who try to stretch as far as others may injure themselves.

Stretching Techniques

Using good stretching techniques can improve flexibility. There are four commonly recognized categories of stretching techniques: static, passive, proprioceptive neuromuscular facilitation (PNF), and ballistic. These are described here and shown later in this chapter.

STATIC STRETCHING

Static stretching involves the gradual lengthening of muscles and tendons as a body part moves around a joint. It is a safe and effective method for improving flexibility. The soldier assumes each stretching position slowly until he feels tension or tightness. This lengthens the muscles without causing a reflex contraction in the stretched muscles. He should hold each stretch for ten seconds or longer. This lets the lengthened muscles adjust to the stretch without causing injury.

The longer a stretch is held, the easier it is for the muscle to adapt to that length. Static stretching should not be painful. The soldier should feel slight discomfort, but no pain. When pain results from stretching, it is a signal that he is stretching a muscle or tendon too much and may be causing damage.

PASSIVE STRETCHING

Passive stretching involves the soldier's use of a partner or equipment, such as a towel, pole, or rubber tubing, to help him stretch. This produces a safe stretch through a range of motion he could not achieve without help. He should talk with his partner to ensure that each muscle is stretched safely through the entire range of motion.

PNF STRETCHING

PNF stretching uses the neuromuscular patterns of each muscle group to help improve flexibility. The soldier performs a series of intense contractions and relaxations using a partner or equipment to help him stretch. The PNF technique allows for greater muscle relaxation following each contraction and increases the soldier's ability to stretch through a greater range of motion.

BALLISTIC STRETCHING

Ballistic, or dynamic, stretching involves movements such as bouncing or bobbing to attain a greater range of motion and stretch. Although this method may improve flexibility, it often forces a muscle to stretch too far and may result in an injury. Individuals and units should not use ballistic stretching.

FITT Factors

Commanders should include stretching exercises in all physical fitness programs.

The following FITT factors apply when developing a flexibility program.

Frequency: Do flexibility exercises daily. Do them during the warm-up to help prepare the muscles for vigorous activity and to help reduce injury. Do them during the cool-down to help maintain flexibility.

Intensity: Stretch a muscle beyond its normal length to the point of tension or slight discomfort, not pain.

Time: Hold stretches for 10 to 15 seconds for warming up and cooling down and for 30 seconds or longer to improve flexibility.

Type: Use static stretches, assumed slowly and gradually, as well as passive stretching and/or PNF stretching.

Warm-Up and Cool-Down

The warm-up and cool-down are very important parts of a physical training session, and stretching exercises should be a major part of both.

THE WARM-UP

The warm-up warms the muscles, increasing the flow of blood and reducing the risk of injury.

Before beginning any vigorous physical activity, one should prepare the body for exercise. The warm-up increases the flow of blood to the muscles and tendons, thus helping reduce the risk of injury. It also increases the joint's range of motion and positively affects the speed of muscular contraction.

A recommended sequence of warm-up activities follows. Soldiers should do these for five to seven minutes before vigorous exercise.

- Slow joggin-in-place or walking for one to two minutes. This causes a gradual increase in the heart rate, blood pressure, circulation, and increases the temperature of the active muscles.
- Slow joint rotation exercises (for example, arm circles, knee/ankle rotations) to gradually increase the joint's range of motion. Work each major joint for 5 to 10 seconds.
- Slow, static stretching of the muscles to be used during the upcoming activity. This will "loosen up" muscles and tendons so they can achieve greater ranges of motion with less risk of injury. Hold each stretch position for 10 to 15 seconds, and do not bounce or bob.
- Calisthenic exerciese, as described in Chapter 7, to increase the intensity level before the activity or conditioning period.
- Slowly mimic the activities to be performed. For example, lift a lighter weight to warm-up before lifting a heavier one. This helps prepare the neuromuscular pathways.

THE COOL-DOWN

The following information explains the importance of cooling down and how to do it correctly.

- Do not stop suddenly after vigorous exercise, as this can be very dangerous. Gradually bring the body back to its resting state by slowly decreasing the intensity of the activity. After running, for example, one should walk for one to two minutes. Stopping exercise suddenly can cause blood to pool in the muscles, thereby reducing blood flow to the heart and brain. This may cause fainting or abnormal rhythms in the heart which could lead to serious complications.
- Repeat the stretches done in the warm-up to help ease muscle tension and any immediate feeling of muscle soreness. Be careful not to overstretch. The muscles are warm from activity and can possibly be overstretched to the point of injury.
- Hold stretches 30 seconds or more during the cool-down to improve flexiblity. Use partner-assisted or PNF techniques, if possible.

The soldier should not limit flexibility training to just the warm-up and cool-down periods. He should sometimes use an entire PT session on a "recovery" or "easy"training day to work on flexibility improvement. He may also work on it at home. Stretching is one form of exercise that takes very little time relative to the benefits gained.

Rotation Exercises

Rotation exercises are used to gently stretch the tendons, ligments, and muscles associated with a joint and to stimulate lubrication of the joint with synovial fluid. This may provide better movement and less friction in the joint.

The following exercises should be performed slowly.

NECK

Position: Stand with the back straight and feet shoulder width apart. Place the hands on the hips.

Action: Roll the head slowly to the left, making a complete circle with the path of the head. Do this three times in each direction.

ARMS AND SHOULDERS

Position: Stand with the back straight and feet shoulder width apart. Extend the arms outward to shoulder height.

Action: Rotate the shoulders forward, and make a large circular motion with the arms. Repeat the action in the opposite direction. Do this three times in each direction.

HIPS

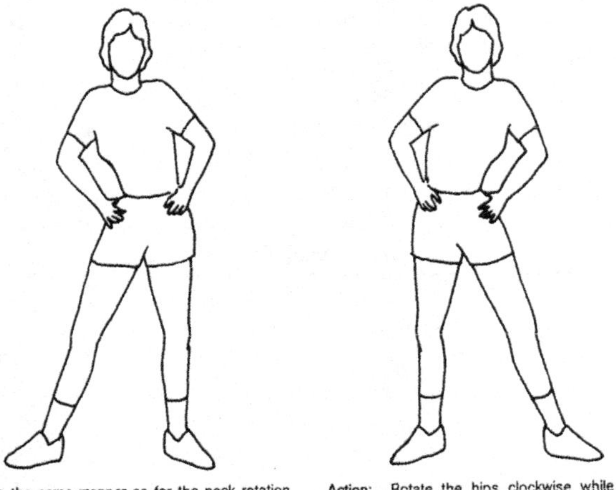

Position: Stand in the same manner as for the neck rotation.

Action: Rotate the hips clockwise while keeping the back straight. Repeat the action in a counterclockwise direction. Do this three times in each direction.

KNEES AND ANKLES

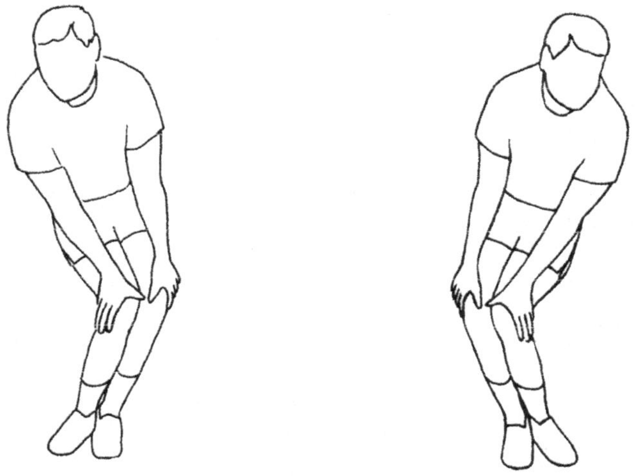

Position: Stand with the feet together, and bend at the waist with the knees slightly bent.

Actio Place the hands above the knees, and rotate the legs in a clockwise direction. Repeat the action in a counterclockwise direction. Do this three times in each direction.

Common Stretching Exercises

The following exercises improve flexibility when performed slowly, regularly, and with gradual progression. Static, passive and PNF stretches are shown.

CAUTION Some of these exercises may be difficult or too strenuous for unfit or medically limited soldiers. Common sense should be used ;n selecting stretching exercises.

STATIC STRETCHES

Assume all stretching positions slowly until you feel tension or slight discomfort. Hold each position for at least 10 to 15 seconds during the warm-up and cool-down. Developmental stretching to improve flexibility requires holding each stretch for 30 seconds or longer.

Choose the appropriate stretch for the muscle groups which you will be working.

NECK AND SHOULDER STRETCH

This stretches the sternocleidomastoid, pectoralis major, and deltoid muscles.

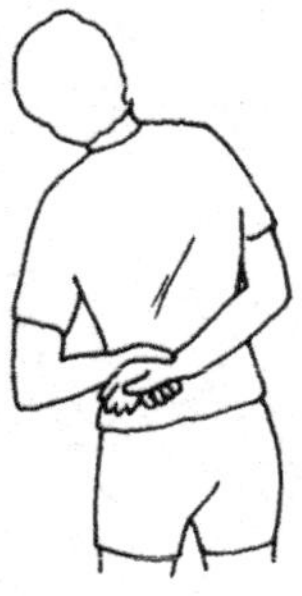

Position: Stand with the feet shoulder width apart and the arms behind the body.

Action: Grasp the left wrist with the right hand. Pull the left arm down and to the right. Tilt the head to the right. Hold this position for 10 to 15 seconds. Repeat the action with the right wrist, pulling the right arm down and to the left. Tilt the head to the left.

ABDOMINAL STRETCH

This stretches the abdominals, obliques, latissimus dorsi, and biceps.

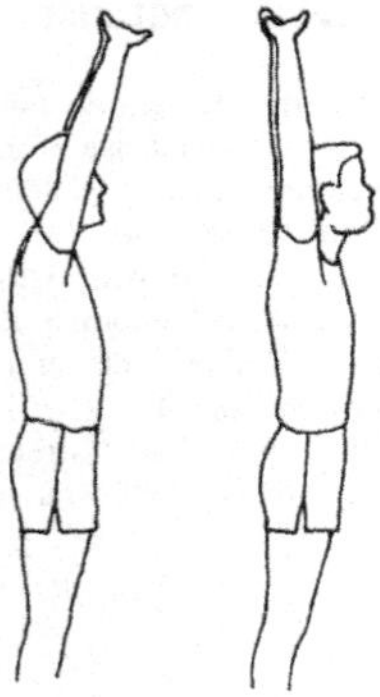

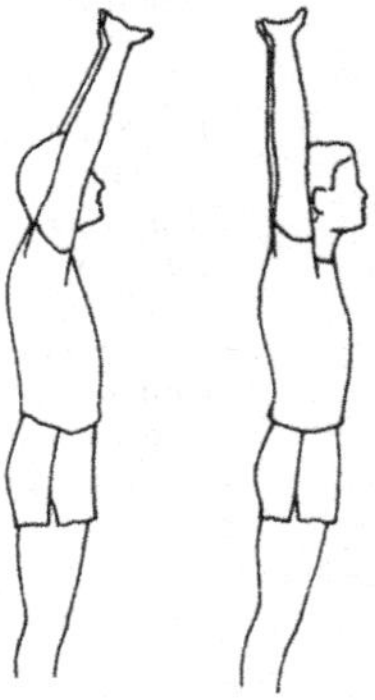

Position: Stand and extend the arms upward and over the head. Interlace the fingers with the palms turned upward.

Action: Stretch the arms up and slightly back. Hold this position for 10 to 15 seconds.
Variation: This stretches the rectus abdominis muscles. Stretch to one side, then the other. Return to the starting position.

CHEST STRETCH

This stretches the pectoralis major, deltoids, and biceps muscle groups.

Position: Stand and interlace the fingers behind the back.

Action: Lift the arms behind the back so that they move outward and away from the body. Lean forward from the waist. Hold this position for 10 to 15 seconds. Bend the knees before moving to the upright position. Return to the starting position.

UPPER-BACK STRETCH

This stretches the lower trapezius and posterior deltoid muscles of the upper back.

Position: Stand with the arms extended to the front at shoulder height with the fingers interlaced and palms facing outward.

Action: Extend the arms and shoulders forward. Hold this position for 10 to 15 seconds. Return to the starting position.

OVERHEAD ARM PULL

This stretches the external and internal obliques, latissimus dorsi, and triceps.

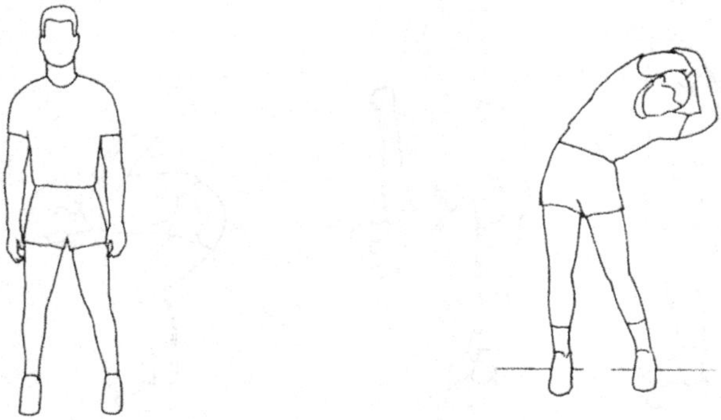

Position: Stand with the feet shoulder width apart. Raise the right arm, bending the right elbow and touching the right hand to the back of the neck.

Action: Grab the right elbow with the left hand, and pull to the left. Hold this position for 10 to 15 seconds. Return to the starting position. Do the same stretch, and pull the left elbow with the right hand for 10 to 15 seconds.

THIGH STRETCH

This stretches the quadriceps and anterior tibialis.

Position: Stand. (For variation, lie on the stomach.)

Action: Bend the left leg up toward the buttocks. Grasp the toes of the left foot with the right hand, and pull the heel to the left buttock. Extend the left arm to the side for balance. Hold this position for 10 to 15 seconds. Return to the starting position. Bend the right leg, grasp the toes of the right foot with the left hand, and pull the heel to the right buttock. Extend the right arm for balance. Hold this position for 10 to 15 seconds. Return to the starting position.

HAMSTRING STRETCH (STANDING)

This stretches the hamstrings, erector spinae, and gluteal muscles.

Position: Stand with the knees slightly bent.

Action: Bend forward keeping the head up, and reach toward the toes. Straighten the legs, and hold this position for 10 to 15 seconds.

HAMSTRING STRETCH (SEATED)

In addition to the muscles mentioned in the standing hamstring stretch, this stretches the calf (gastrocnemius and soleus) muscles.

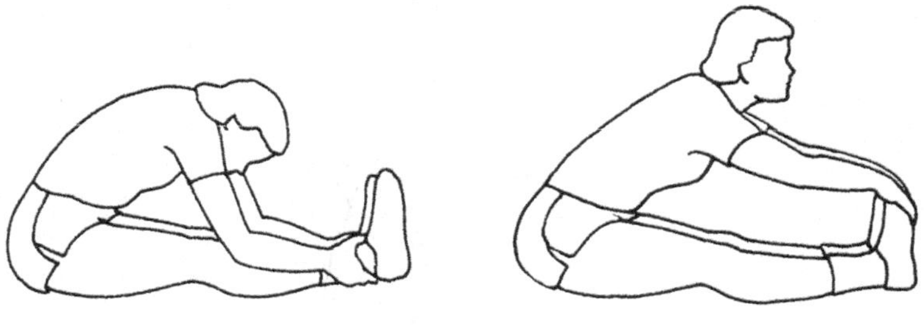

Position: Sit on the ground with both legs straight and extended forward with the feet upright about six inches apart. Put the hands on the ankles or toes.

Action: Bend from the hips, keeping the back and head in a comfortable, straight line. Hold this position for 10 to 15 seconds. (Variation for greater stretch: Stretch and pull back on the toes.)

GROIN STRETCH (STANDING)

This stretches the hip adductor muscles.

Position: Lunge slowly to the left while keeping the right leg straight, the right foot facing straight ahead and entirely on the floor.

Action: Lean over the left leg while stretching the right groin muscles. Hold this position for 10 to 15 seconds. Repeat with the opposite leg.

GROIN STRETCH (SEATED)

This stretches the hip adductor and erector spinae muscles.

Position: Sit on the ground with the soles together. Place the hands on or near the feet.

Action: Bend forward from the hips, keeping the head up. Hold this position for 10 to 15 seconds.

GROIN STRETCH (SEATED STRADDLE)

This stretches the hip adductor (on the inside of the upper leg), gluteals, erector spinae, and hamstring muscles.

Position: Sit on the ground with the legs straight and spread as far apart as possible.

Action: Bend forward at the hips, keep the head up, and reach toward the feet. Hold this position for 10 to 15 seconds. Variation: Stretch to one side while trying to touch the toes. Next, stretch to the other side.

CALF STRETCH

This stretches the calf (gastrocnemius and soleus) muscles.

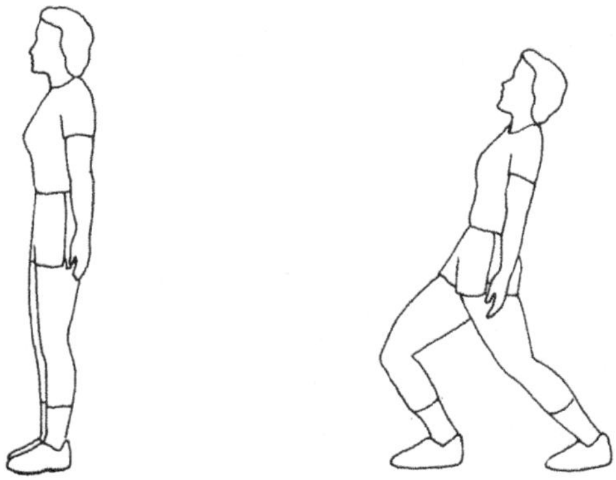

Position: Stand straight with the feet together, arms extended downward, elbows locked, palms facing backward, fingers extended and joined, and head and eyes facing front.

Action: Move the right foot to the rear about two feet, and place the ball of the foot on the ground. Slowly press the right heel to the ground. Slowly bend the left knee while pushing the hips forward and arching the back slightly. Hold this position for 10 to 15 seconds. Return to the starting position. Repeat with the left foot. Return to the starting position.

CALF STRETCH (VARIATION: TOE PULL)

This stretches the calf (gastrocnemius) and to a lesser extent the hamstrings, gluteus maximus, and erector spinae muscles.

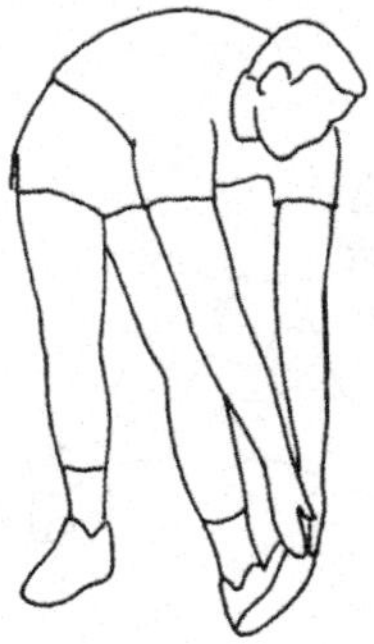

Position: Stand with the feet shoulder width apart and the left foot slightly forward.

Action: Bend forward at the waist. Slightly bend the right knee, and fully extend the left leg. Reach down and pull the toes of the left foot toward the left shin. Hold this position for 10 to 15 seconds. Return to the starting position. In a similar manner, pull the toes of the right foot toward the right shin, and hold for 10 to 15 seconds.

HIP AND BACK STRETCH (SEATED)

This stretches the hip abductors, erector spinae, latissimus dorsi, and oblique muscle groups.

Position: Sit on the ground with the right leg forward and straight. Cross the left leg over the right while sitting erect. Keep the heels of both feet in contact with the ground.

Action: Slowly rotate the upper body to the left and look over the left shoulder. Reach across the left leg with the right arm, and push the left leg to your right. Use the left hand for support by placing it on the ground. Hold this position for 10 to 15 seconds. Repeat this stretch for the other side by crossing and turning in the opposite direction.

HIP AND BACK STRETCH (LYING DOWN)

This stretches the gluteal and erector spinae muscles.

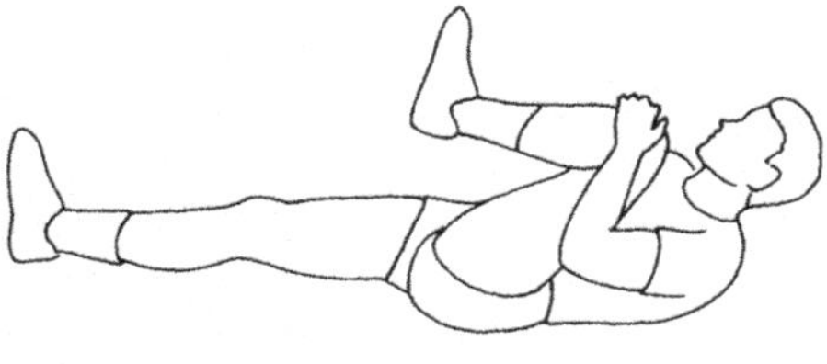

Action 1

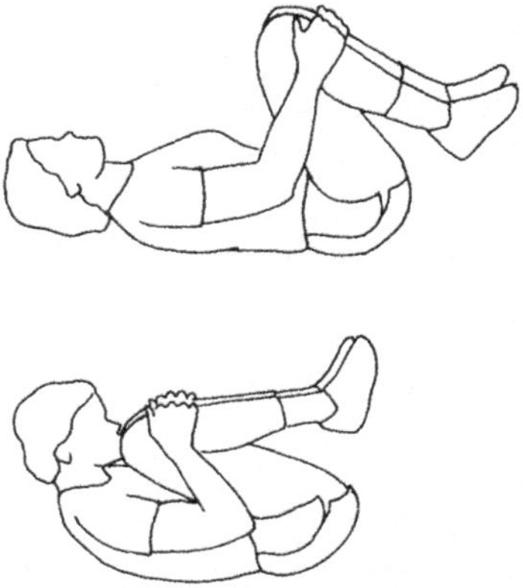

Action 2

Position: Lie on the back with the arms straight beside the body. Keep the legs straight and the knees and feet together.
Action 1: Bring the left leg straight back toward the head, leaving the right leg in the starting position. Bring the head and arms up. Grab the bent left leg below the knee, and pull it gradually to the chest. Hold this position for 10 to 15 seconds. Gradually return to the starting position. Repeat these motions with the opposite leg.

Action 2: *Pull both knees to the chest.* Pull the head up to the knees. Hold for 10 to 15 seconds. Return to the starting position.

PASSIVE STRETCHES

Passive stretching is done with the help of a partner or equipment. The examples in this chapter show passive stretching done with a towel or with a partner. When stretching alone, using a towel may help the exerciser achieve a greater range of motion.

TOWEL STRETCHES

1

This stretches the abdominal and pectoral muscles.
Position: Stand erect with the hands overhead and grasping a towel.
Action: Pull tightly on the towel while reaching up and slightly arching the back. Hold for 10 to 15 seconds.

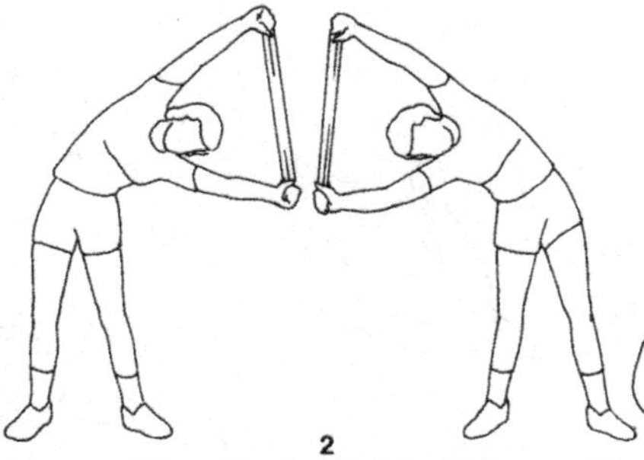

2

This stretches the abdominals, obliques, and latissimus dorsi.
Position: Stand erect with the hands overhead and grasping a towel.
Action: Slowly bend sideways to the left as far as possible. Hold for 10 to 15 seconds. Repeat for the opposite side. While doing this stretch, pulling on the towel with the bottom arm will enhance the stretch.

3

This stretches the hamstring, calf, and low back muscles.
Position: Sit with the legs straight and together. Grasping each end of a short towel, place the middle of the towel over the balls of the feet.
Action: Pulling on the towel, come forward as far as possible keeping the legs straight and the toes pulled back.

PARTNER-ASSISTED CHEST STRETCH

This exercise stretches the pectoralis major, deltoids, and biceps muscles.

Position: Sit erect with the arms straight, elevated to shoulder height, and the palms facing forward. The partner stands behind the exerciser grasping the arms between the wrists and the elbows.

Action: The partner gradually pulls both of the exerciser's arms toward the rear until the stretch causes the exerciser mild discomfort. Hold this position for 10 to 15 seconds.

PARTNER-ASSISTED HAMSTRING STRETCH

This exercise stretches the hamstrings and erector spinae muscle groups.

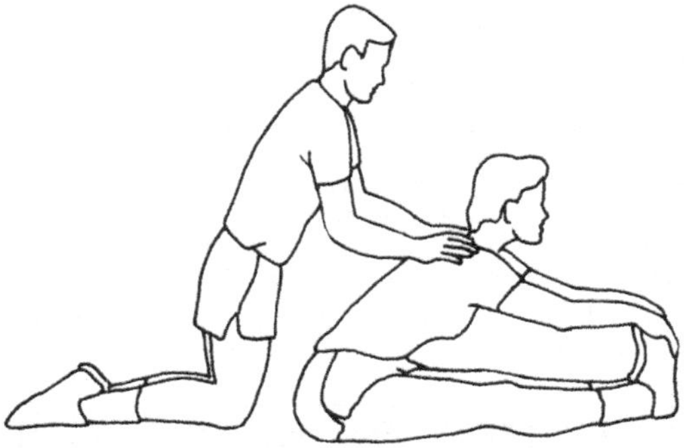

Position: Sit erect on the ground with the legs together. The partner kneels behind the exerciser. If the partner stands, he may apply too much pressure.

Action: The partner places light pressure on the exerciser's upper back until the exerciser's forward motion results in mild discomfort. This position is held for 10 to 15 seconds.

PARTNER-ASSISTED GROIN STRETCH

This exercise stretches the hip adductor and erector spinae muscle groups.

Position: Sit on the ground with knees bent and soles together. The partner kneels behind the exerciser. If the partner stands, he may apply too much pressure.

Action: The partner places light pressure on the exerciser's knees with his hands and leans gently on the exerciser's back with his chest until the stretch causes the exerciser mild discomfort. This position is held for 10 to 15 seconds.

Soldiers can do PNF (Proprioceptive Neuromuscular Facilitation) stretches for most major muscle groups. PNF stretches use a series of contractions, done against a partner's resistance, and relaxations.

Obtaining a safe stretch beyond the muscle's normal length requires a partner's assistance. The following four steps provide general guidance as to how PNF stretches are done. Both the exerciser and partner should follow these instructions:

1. Assume the stretch position slowly with the partner's help.
2. Isometrically contract the muscles to be stretched. Hold the contraction for 5 to 10 seconds against the partner's unyielding resistance.
3. Relax. Next, contract the antagonistic muscles for 5 to 10 seconds while the partner helps the exerciser obtain a greater stretch.
4. Repeat this sequence three times, and try to stretch a little further each time. (Caution: The exerciser should not hold his breath. He should breathe out during each contraction.)

Several examples of PNF stretches are provided below in a stepwise fashion. The numbers given above for each step correspond to the general description listed below.

PNF HAMSTRING AND GLUTEAL STRETCH

1. The exerciser lies on his back and places the lower part of his left leg on the partner's right shoulder. The exerciser slowly stretches the hamstring and gluteal muscles by gradually bringing the straightened leg toward his head until he feels tension in the stretched muscles. The partner then applies light pressure on the exerciser's lower leg to help maintain or further the stretch.
2. The exerciser isometrically contracts his hamstring and gluteal muscles for 5 to 10 seconds by trying to move his leg downward and away from his head. The partner steadily resists the exerciser's efforts and does not allow any movement to occur.
3. The exerciser relaxes the hamstring and gluteal muscles. He then tries to stretch them farther by using the partner's help and by contracting the antagonistic, hip flexor muscles (the iliopsoas and quadriceps) and the tibialis anterior muscle for 5 to 10 seconds.
4. Perform these movements three times for each leg. Try to stretch a little further each time.

PNF CHEST STRETCH

1. The exerciser and partner assume the positions used in the partner-assisted chest stretch. The exerciser slowly stretches until he feels tension in the stretched muscles. The partner then applies light pressure on the forearm to help maintain or further the stretch.

2. The exerciser contracts the pectorals for 5 to 10 seconds, trying to pull his arms and hands together against the partner's unyielding resistance.

3. The exerciser relaxes his pectorals and attempts to stretch further with the partner's help while contracting the antagonistic muscles of the upper back. He does this for 5 to 10 seconds.

4. Repeat this sequence three times, trying to stretch a little further each time.

PNF GROIN STRETCH

1. The exerciser assumes the position used for the partner-assisted groin stretch. The partner kneels behind him. The exerciser slowly lowers his legs and leans forward until tension is felt in the muscles of the groin (his abductors) and lower back (erector spinae muscles). Next, the partner applies light pressure on the exerciser's thighs and back to help maintain or further increase the stretch.

2. The exerciser then attempts to push upward for 5 to 10 seconds by contracting the groin and lower back muscles while the partner resists and allows no movement to occur.

3. The exerciser relaxes the groin and lower back muscles and tries to stretch further with the partner's help and by contracting the antagonistic muscles (hip abductor and abdominal muscles) for 5 to 10 seconds.

4. Perform these movements three times. Try to stretch a little further each time.

CHAPTER 5

Body Composition

Body composition, which refers to the body's relative amounts of fat and lean body mass (organs, bones, muscles), is one of the five components of physical fitness. Good body composition is best gained through proper diet and exercise. Examples of poor body composition are underdeveloped musculature or excessive body fat. Being overweight (that is, overly fat) is the more common problem.

Poor body composition causes problems for the Army. Soldiers with inadequate muscle development cannot perform as well as soldiers with good body composition. As a soldier gets fat, his ability to perform physically declines, and his risk of developing disease increases. Soldiers with high percentages of body fat often have lower APFT scores than those with lower percentages. Poor body composition, especially obesity, has a negative effect on appearance, self--esteem, and negatively influences attitude and morale.

The Army's weight control program is described in AR 600-9. It addresses body composition standards, programs for the overly fat, and related administrative actions.

The amount of fat on the body, when expressed as a percentage of total body weight, is referred to as the percent body fat. The Army's maximum allowable percentages of body fat, by age and sex, are listed in Figure 5-1.

Evaluation Methods

The Army determines body fat percentage using the girth method. (This is described in AR 600-9, pages 12 to 21.)

Body composition is influenced by age, diet, fitness level, and genetic factors (gender and body type). The Army's screening charts for height and weight (shown in AR 600-9) make allowances for these differences. A soldier whose weight exceeds the standard weight shown on the charts may not necessarily be overfat. For example, some well-muscled athletes have body weights that far exceed the values for weight listed on the charts for their age, gender, and height. Yet, only a small percentage of their total body mass may be fat. In such cases, the lean body mass accounts for a large share of their total body composition, while only a small percentage of the total body mass is composed of fat.

Body composition is influenced by age, fitness level, and genetic factors.

Soldiers who do not meet the weight standards for their height and/or soldiers whose appearance suggests that they have excessive fat are to be evaluated using the circumference (girth measurement) method described in AR 600-9.

BODY FAT STANDARDS

AGES:	17-20	21-27	28-39	40+
MALES 20%	20%	22%	24%	26%
FEMALES	30%	32%	34%	36%

Figure 5-1

A more accurate way to determine body composition is by hydrostatic or underwater weighing. However, this method is very time-consuming and expensive and usually done only at hospitals and universities.

Soldiers who do not meet Army body fat standards are placed on formal, supervised weight (fat) loss programs as stipulated in AR 600-9. Such programs include sensible diet and exercise regimens.

Diet and Exercise

A combination of exercise and diet is the best way to lose unwanted body fat.

A combination of exercise and diet is the best way to lose excessive body fat. Losing one to two pounds a week is a realistic goal which is best accomplished by reducing caloric intake and increasing energy expenditure. In other words, one should eat less and exercise more. Dieting alone can cause the body to believe it is being starved. In response, it tries to conserve its fat reserves by slowing down its metabolic rate and, as a result, it loses fat at a slower rate.

Soldiers must consume a minimum number of calories from all the major food groups, with the calories distributed over all the daily meals including snacks. This ensures an adequate consumption of necessary vitamins and minerals. A male soldier who is not under medical supervision when dieting requires a caloric intake of at least 1 ,500; women require at least 1,200 calories. Soldiers should avoid diets that fail to meet these criteria.

Aerobic exercise is best for burning fat. examples include jogging, walking, swimming, bicycling, cross-country skiing, and rowing.

Trying to lose weight with fad diets and devices or by skipping meals does not work for long-term fat loss, since weight lost through these practices is mostly water and lean muscle tissue, not fat. Losing fat safely takes time and patience. There is no quick and easy way to improve body composition.

The soldier who diets and does not exercise loses not only fat but muscle tissue as well. This can negatively affect his physical readiness. Not only does exercise burn calories, it helps the body maintain its useful muscle mass, and it may also help keep the body's metabolic rate high during dieting.

Fat can only be burned during exercise if oxygen is used. Aerobic exercise, which uses lots of oxygen, is the best type of activity for burning fat. Aerobic exercises include jogging, walking, swimming, bicycling, cross-country skiing, rowing, stair climbing, exercise to music, and jumping rope. Anaerobic activities, such as sprinting or lifting heavy weights, burn little, if any, fat.

Exercise alone is not the best way to lose body fat, especially in large amounts. For an average-sized person, running or walking one mile burns about 100 calories. Because there are 3,500 calories in one pound of fat, he needs to run or walk 35 miles if pure fat were being burned. In reality, fat is seldom the only source of energy used during aerobic exercise. Instead, a mixture of both fats and carbohydrates is used. As a result, most people would need to run or walk over 50 miles to burn one pound of fat.

A combination of proper diet and aerobic exercise is the proven way to lose excessive body fat. Local dietitians and nutritionists can help soldiers who want to lose weight by suggesting safe and sensible diet programs. In addition, the unit's MFT can design tailored exercise programs which will help soldiers increase their caloric expenditure and maintain their lean body mass.

CHAPTER 6

Nutrition and Fitness

In addition to exercise, proper nutrition plays a major role in attaining and maintaining total fitness. Good dietary habits (see Figure 6-1) greatly enhance the ability of soldiers to perform at their maximum potential. A good diet alone, however, will not make up for poor health and exercise habits. This chapter gives basic nutritional guidance for enhancing physical performance. Soldiers must know and follow the basic nutrition principles if they hope to maintain weight control as well as achieve maximum physical fitness, good health, and mental alertness.

Guidelines for Healthy Eating

Eating a variety of foods and maintaining an energy balance are basic guidelines for a healthy diet. Good nutrition is not complicated for those who understand these dietary guidelines.

To be properly nourished, soldiers should regularly eat a wide variety of foods fro-m the major food groups, selecting a variety of foods from within each group. (See Figure 6-2.) A well-balanced diet provides all the nutrients needed to keep one healthy.

Most healthy adults do not need vitamin or mineral supplements if they eat a proper variety of foods. There are no known advantages in consuming excessive amounts of any nutrient, and there may be risks in doing so.

For soldiers to get enough fuel from the food they eat and to obtain the variety of foods needed for nutrient balance, they should eat three meals a day. Even snacking between meals can contribute to good nutrition if the right foods are eaten.

Another dietary guideline is to consume enough calories to meet one's energy needs. Weight is maintained as long as the body is in energy balance,

DIETARY GUIDELINES

- **Eat a Variety of Foods**
- **Maintain a Healthy Body Weight**
- **Choose a Diet Low in Fat, Saturated Fat, and Cholesterol**
- **Choose a Diet with Plenty of Vegetables, Fruits, and Grain Products**
- **Use Sugars Only in Moderation**
- **Use Salt and Sodium Only in Moderation**
- **If you Drink Alcoholic Beverages, Do So in Moderation**

Figure 6-1

DAILY FOOD GUIDE		
Eat a variety of foods from each food group. Most people should have the minimum number of servings; others need more due to their body size and activity level.		
FOOD GROUP	**SUGGESTED NUMBER OF SERVINGS**	**SUGGESTED SIZE OF SERVINGS**
Vegetables (Include dark green, leafy, or deep yellow ones)	3 to 5	1 cup of raw, leafy greens or 1/2 cup of cooked vegetables
Fruits (Include citrus fruits or juices, melons, or berries)	2 to 4	1 medium fruit or 1/2 cup of diced or small fruit or 3/4 cup of juice
Breads, Cereals, Rice, and Pasta (Include whole grain varieties)	6 to 11	1 slice of bread, 1/2 bun or roll, 1/2 cup of cooked cereal, rice or pasta, 1 oz. of ready-to- eat cereal
Milk, Yogurt, and Cheese (Include skim or lowfat varieties)	2 to 3	1 cup of milk or yogurt, 1-1/2 oz. of hard cheese
Meats, Poultry, Fish, Dry Beans or Peas, Eggs, Nuts (Use lean meats and remove skin from poultry)	2 to 3	2 or 3 oz. of cooked meat, fish, or poultry (TOTAL 6 oz/day) 2 eggs, or 1 cup of cooked beans or peas

Figure 6-2

that is, when the number of calories used equals the number of calories consumed.

The most accurate way to control caloric intake is to control the size of food portions and thus the total amount of food ingested. One can use standard household measuring utensils and a small kitchen scale to measure portions of foods and beverages. Keeping a daily record of all foods eaten and physical activity done is also helpful.

Figure 6-3 shows the number of calories burned during exercise periods of different types, intensities, and durations. For example, while participating in archery, a person will burn 0.034 calories per pound per minute. Thus, a 150-pound person would burn 5.1 calories per minute (150 lbs. x 0.034 calories/minute/lb. = 5.1 calories/minute) or about 305 calories/hour, as shown in Figure 6-4. Similarly, a person running at 6 miles per hour (MPH) will burn 0.079 cal./min./lb. and a typical, 150-pound male will burn 11.85 calories/minute (150 lbs. x 0.079 cal./lb./min. = 11.85) or about 710 calories in one hour, as shown in Figure 6-3.

To estimate the number of calories you use in normal daily activity, multiply your body weight by 13 if you are sedentary, 14 if somewhat active, and 15 if moderately active. The result is a rough estimate of the number of calories you need to maintain your present body weight. You will need still more calories if you are more than moderately active. By comparing caloric intake with caloric expenditure, the state of energy balance (positive, balanced, or negative) can be determined.

CALORIC EXPENDITURE CHART

ACTIVITY	CAL/MIN/LB	CAL/HR/150 LB
Archery	.034	305
Badminton:		
Moderate	.039	350
Vigorous	.065	585
Basketball:		
Moderate	.047	420
Vigorous	.066	595
Baseball:		
Infield-outfield	.031	280
Pitching	.039	350
Bicycling:		
Slow (5 MPH)	.025	225
Moderate (10 MPH)	.05	450
Fast (13 MPH)	.072	650
Bowling	.028	255
Calisthenics:		
General	.045	405
Canoeing:		
2.5 MPH	.023	210
4.0 MPH	.047	420
Dancing:		
Slow	.029	260
Moderate	.045	405
Fast	.064	575
Fencing:		
Moderate	.033	300
Vigorous	.057	515
Fishing	.016	145
Football (tag)	.04	360
Gardening	.024	220
Gardening-Weeding	.039	260
Golf	.029	260
Gymnastics:		
Light	.022	200
Heavy	.056	505
Handball	.063	570
Hiking	.042	375
Hill Climbing	.06	540
Hoeing, Raking, Planting	.031	280
Horseback Riding:		
Walk	.019	175
Trot	.046	415
Gallop	.067	600
Jogging:		
4.5 MPH (13:330 mi.)	.063	565

ACTIVITY	CAL/MIN/LB	CAL/HR/150 LB *
Judo, Karate	.087	785
Motor Boating	.016	145
Mountain Climbing	.086	775
Rowing		
(Rec 2.5 MPH)	.036	325
Vigorous	.118	1000
Running:		
6 MPH (10 min/mi)	.079	710
10 MPH (6 min/mi)	.1	900
12 MPH (5 min/mi)	.13	1170
Sailing	.02	180
Skating:		
Moderate (Rec)	.036	325
Vigorous	.064	575
Skiing (Snow):		
Downhill	.059	530
Level (5 MPH)	.078	700
Soccer	.06	570
Squash	.07	630
Stationary Run:		
70-80 cts/min	.078	705
Strength Training (10 rep circuit)		
60% 1RM	.022	198
80% 1RM	.048	432
Swimming (crawl):		
20 yds/min	.032	290
45 yds/min	.058	520
50 yds/min	.071	640
Table Tennis:		
Moderate	.026	235
Vigorous	.06	540
Tennis:		
Moderate	.046	415
Vigorous	.04	540
Volleyball:		
Moderate	.036	325
Vigorous	.065	585
Walking:		
2.0 MPH	.022	200
3.0 MPH	.03	270
4.0 MPH	.039	350
5.0 MPH	.064	575
Water Skiing	.053	480
Wrestling	.091	820

* A 150-pound person will expend the number of calories indicated in one hour for any given activity.

Figure 6-3

Avoiding an excessive intake of fats is an important fundamental of nutrition.

Avoiding an excessive intake of fats is another fundamental dietary guideline. A high intake of fats, especially saturated fats and cholesterol, has been associated with high levels of blood cholesterol.

The blood cholesterol level in most Americans is too high. Blood cholesterol levels can be lowered by reducing both body fat and the amount of fat in the diet. Lowering elevated blood cholesterol levels reduces the risk of developing coronary artery disease (CAD) and of having a heart attack. CAD, a slow, progressive disease, results from the clogging of blood vessels in the heart. Good dietary habits help reduce the likelihood of developing CAD.

Carbohydrates are the primary fuel source for muscles during short-term, high-intensity activities.

It is recommended that all persons over the age of two should reduce their fat intake to 30 percent or less of their total caloric intake. The current national average is 38 percent. In addition, we should reduce our intake of saturated fat to less than 10 percent of the total calories consumed. We should increase our intake of polyunsaturated fat, but to no more than 10 percent of our total calories. Finally, we should reduce our daily cholesterol intake to 300 milligrams or less. Figure 6-4 suggests actions commanders can take to support sound dietary guidelines. Most of these actions concern dining-facility management.

Concerns for Optimal Physical Performance

Carbohydrates, in the form of glycogen (a complex sugar), are the primary fuel source for muscles during short-term, high-intensity activities. Repetitive, vigorous activity can use up most of the carbohydrate stores in the exercised muscles.

The body uses fat to help provide energy for extended activities such as a one-hour run. Initially, the chief fuel burned is carbohydrates, 'but as the duration increases, the contribution from fat gradually increases.

The intensity of the exercise also influences whether fats or carbohydrates are used to provide energy. Very intense activities use more carbohydrates. Examples include weight training and the APFT sit-up and push-up events.

Eating foods rich in carbohydrates helps maintain adequate muscle-glycogen reserves while sparing amino acids (critical building-blocks needed for building proteins). At least 50 percent of the calories in the diet should come from carbohydrates. Individual caloric requirements vary, depending on body size, sex, age, and training mission. Foods rich in complex carbohydrates (for example, pasta, rice, whole wheat bread, potatoes) are the best sources of energy for active soldiers.

COMMANDER'S CHECKLIST FOR NUTRITION

PRINCIPLES OF NUTRITION	SUPPORTING ACTIONS
1. Eat a variety of foods. No single food item provides all essential nutrients.	In the dining facility: • Ensure menus provide foods from the basic 4 food groups: fruits and vegetables, meats, dairy products, and breads and cereals. • Establish serving lines in the following order, if possible: (1) salads, (2) fruits, (3) entrees, (4) hot vegetables, (5) breads, (6) beverages, (7) desserts.
2. Maintain a desirable body weight. Excess body fat detracts from fitness. Weight loss is achieved by increasing physical activity and decreasing total food intake, especially fats, refined sugars, and alcohol.	In the dining facility, provide: • Low-calorie menu, including short-order items at each meal. Use the Master Menu (SB 10-260) menu patterns. • Reduced-portion sizes. • No-calorie beverages. • Low-calorie salad dressings. • Posted list of caloric values of menu items, before or on the serving line.
3. Avoid excess dietary fat. Too much fat (especially cholesterol and saturated fat) can lead to heart disease and weight problems. Fats contain twice as many calories as equal amounts of carbohydrates or protein.	In the dining facility, provide: • Non-fried eggs as an alternative. • Margarine as a butter alternative. • Two percent milk as the primary milk in bulk dispensers. • Skim milk in 1/2-pint cartons. • Sauces, gravies, and margarine separately from the entree or vegetable. • Avoid animal fats, palm oil, and hydrogenated vegetable oil.
4. Avoid too much sugar. Sweets are empty calories and may lead to dental cavities and weight problems.	In the dining facility, provide: • Fruit as a dessert alternative. • Unsweetened juices. • No-calorie, unsweetened beverages. • Non-nutritive, sugar substitute as a granulated sugar alternative. • Unsweetened cereal.
5. Eat foods with adequate starch and fiber. Eating complex carbohydrates adds to the diet and reduces symptoms of constipation.	In the dining facility, provide: • Whole-grain breads, cereals and legumes. • Fresh fruit. • Salad bars at lunch and dinner.
6. Avoid too much sodium. Eating highly-salted foods may lead to excessive sodium intake. This may be a problem for those "at risk" for high blood pressure.	• Reduce salt in recipes by 25 percent.
7. If you drink alcoholic beverages, do so in moderation. Alcoholic beverages are high in calories and and low in nutrients. One or two standard-size drinks daily appears to cause no harm in normal, healthy, nonpregnant adults.	• Avoid alcohol; it is detrimental to good health and weight management.
8. Know the nutrition principles. Educating soldiers maximizes efforts to improve nutritional fitness. Reference: AR 30-1, Appendix J.	• Display educational materials on nutrition; (posters, table tents, bulletin boards, and handouts). • Provide food-service personnel with training programs on nutrition standards. • Provide unit-training programs on nutrition for soldiers. (Use installation dietitian).

Figure 6-4

Because foods eaten one to three days before an activity provide part of the fuel for that activity, it is important to eat foods every day that are rich in complex carbohydrates. It is also important to avoid simple sugars, such as candy, up to 60 minutes before exercising, because they can lead to low blood sugar levels during exercise.

Soldiers often fail to drink enough water, especially when training in the heat. Water is an essential nutrient that is critical to optimal physical performance. It plays an important role in maintaining normal body temperature. The evaporation of sweat helps cool the body during exercise. As a result, water lost through sweating must be replaced or poor performance, and possibly injury, can result. Sweat consists primarily of water with small quantities of minerals like sodium. Cool, plain water is the best drink to use to replace the fluid lost as sweat. Soldiers should drink water before, during, and after exercise to prevent dehydration and help enhance performance. Figure 6-5 shows recommendations for fluid intake when exercising.

Sports drinks, which are usually simple carbohydrates (sugars) and electrolytes dissolved in water, are helpful under certain circumstances. There is evidence that solutions containing up to 10 percent carbohydrate will enter the blood fast enough to deliver additional glucose to the active muscles. This can improve endurance.

During prolonged periods of exercise (1.5+ hours) at intensities over **50** percent of heart rate reserve, one can benefit from periodically drinking sports drinks with a concentration of 5 to 10 percent carbohydrate. Soldiers on extended road marches can also benefit from drinking these types of glucose-containing beverages. During intense training, these beverages can provide a source of carbohydrate for working muscles. On the other hand, drinks that exceed levels of 10 percent carbohydrate, as do regular soda pops and most fruit juices, can lead to abdominal cramps, nausea, and diarrhea. Therefore, these drinks should be used with caution during intense endurance training and other similar activities.

Many people believe that body builders need large quantities of

RECOMMENDATIONS FOR FLUID INTAKE

- Drink cool (40 degrees F) water. This is the best drink to sustain performance. Fluid also comes from juice, milk, soup, and other beverages.
- Do not drink coffee, tea, and soft drinks even though they provide fluids. The caffeine in them acts as a diuretic which can increase urine production and fluid loss. Avoid alcohol for the same reason.
- Drink large quantities (20 oz.) of water one or two hours before exercise to promote hyperhydration. This allows time for adequate hydration and urination.
- Drink three to six ounces of fluid every 15 to 30 minutes during exercise.
- Replace fluid sweat losses by monitoring pre-and post-exercise body weights. Drink two cups of fluid for every pound of weight lost.

Figure 6-5

protein to promote better muscle growth. The primary functions of protein are to build and repair body tissue and to form enzymes. Protein is believed to contribute little, if any, to the total energy requirement of heavy-resistance exercises. The recommended dietary allowance of protein for adults is 0.8 grams per kilogram of body weight. Most people meet this level when about 15 percent of their daily caloric intake comes from protein. During periods of intense aerobic training, one's need for protein might be somewhat higher (for example, 1.0 to 1.5 grams per kilogram of body weight per day). Weight lifters, who have a high proportion of lean body mass, can easily meet their protein requirement with a well-balanced diet which has 15 to 20 percent of its calories provided by protein. Recent research suggests that weight trainers may need no more protein per kilogram of body weight than average, nonathletic people. Most Americans routinely consume these levels of protein, or more. The body converts protein consumed in excess of caloric needs to fat and stores it in the body.

Nutrition in the Field

Soldiers in the field must eat enough food to provide them with the energy they need. They must also drink plenty of water or other non-alcoholic beverages. The "meal, ready to eat" (MRE) supplies the needed amount of carbohydrates, protein, fat, vitamins, and minerals. It is a nutritionally adequate ration when all of its components are eaten and adequate amounts of water are consumed. Because the foods are enriched and fortified with vitamins and minerals, each component is a major source of nutrients. Soldiers must eat all the components in order to get the daily military recommended dietary allowances (MRDA) and have an adequate diet in the field. Soldiers who are in weight control programs or who are trying to lose weight can eat part of each MRE item, as recommended by dietitians.

CHAPTER 7

Circuit Training and Exercise Drills

This chapter gives commanders and trainers guidance in designing and using exercise circuits. It describes calisthenic exercises for developing strength, endurance, coordination, and flexibility. It also describes grass drills and guerilla exercises which are closely related to soldiering skills and should be regularly included in the unit's physical fitness program.

Circuit training is a term associated with specific training routines. Commanders with a good understanding of the principles of circuit training may apply them to a wide variety of training situations and environments.

Circuits

A circuit is a group of stations or areas where specific tasks or exercises are performed.

A circuit is a group of stations or areas where specific tasks or exercises are performed. The task or exercise selected for each station and the arrangement of the stations is determined by the objective of the circuit.

Circuits are designed to provide exercise to groups of soldiers at intensities which suit each person's fitness level. Circuits can promote fitness in a broad range of physical and motor fitness areas. These include CR endurance, muscular endurance, strength, flexibility, and speed. Circuits can also be designed to concentrate on sports skills, soldiers' common tasks, or any combination of these. In addition, circuits can be organized to exercise all **the** fitness components in a short period of time. A little imagination can make circuit training an excellent addition to a unit's total physical fitness program. At the same time, it can provide both fun and a challenge to soldiers' physical and mental abilities. Almost any area can be used, and any number of soldiers can exercise for various lengths of time.

TYPES OF CIRCUITS

The two basic types of circuits are the free circuit and the fixed circuit. Each has distinct advantages.

Free Circuit

In a free circuit, there is no set time for staying at each station, and no signal is given to move from one station to the next. Soldiers work at their own pace, doing a fixed number of repetitions at each station. Progress is measured by the time needed to complete a circuit. Because soldiers may do incomplete or fewer repetitions than called for to reduce this time, the quality and number of the repetitions done should be monitored. Aside from this, the free circuit requires little supervision.

Fixed Circuit

In a fixed circuit, a specific length of time is set for each station. The time is monitored with a stopwatch, and soldiers rotate through the stations on command.

There are three basic ways to increase the intensity or difficulty of a fixed circuit:

- Keep the time for completion the same, but increase the number of repetitions.
- Increase the time per station along with the number of repetitions.
- Increase the number of times soldiers go through the circuit.

VARIABLES IN CIRCUIT TRAINING

Several variables in circuit training must be considered. These include the time, number of stations, number of time, number of stations, number of soldiers, number of times the circuit is completed, and sequence of stations. These are discussed below.

Time

One of the first things to consider is how long it should take to complete the circuit. When a fixed circuit is run, the time at each station should always be the same to avoid confusion and help maintain control. Consider also the time it takes to move from one station to the next. Further, allow from five to seven minutes both before and after running a circuit for warming up and cooling down, respectively.

Number of Stations

The objective of the circuit and time and equipment available strongly influence the number of stations. A circuit geared for a limited objective (for example, developing lower-body strength) needs as few as six to eight stations. On the other hand, circuits to develop both strength and CR fitness may have as many as 20 stations.

Number of Soldiers

If there are 10 stations and 40 soldiers to be trained, the soldiers should be divided into 10 groups of four each. Each station must then be equipped to handle four soldiers. For example, in this instance a rope jumping station must have at least four jump ropes. It is vital in a free circuit that no soldier stand around waiting for equipment. Having enough equipment reduces bottlenecks, slowdowns, and poor results.

Number of Times a Circuit is Completed

To achieve the desired training effect, soldiers may have to repeat the same circuit several times. For example, a circuit may have ten stations. Soldiers may run through the circuit three times, exercising for 30 seconds at each station, and taking 15 seconds to move between stations. The exercise time at each station may be reduced to 20 seconds the second and third time through. The whole workout takes less than 45 minutes including warm-up and cool-down. As soldiers become better conditioned, exercise periods may be increased to 30 seconds or longer for all three rotations. Another option is to have four rotations of the circuit.

Sequence of Stations

Stations should be arranged in a sequence that allows soldiers some recovery time after exercising at strenuous stations. Difficult exercises can be alternated with less difficult ones. After the warm-up, soldiers can start a circuit at any station and still achieve the objective by completing the full circuit.

DESIGNING A CIRCUIT

The designer of a circuit must consider many factors. The six steps below cover the most important aspects of circuit development.

Determine Objectives

The designer must consider the specific parts of the body and the components of fitness on which soldiers need to concentrate. For example, increasing muscular strength may be the primary objective, while muscular endurance work may be secondary. On the other hand, improving cardiorespiratory endurance may be the top priority. The designer must first identify the training objective in order to choose the appropriate exercises.

The designer must consider the specific parts of the body and the components of fitness on which soldiers need to concentrate.

Select the Activities

The circuit designer should list all the exercises or activities that can help meet the objectives. Then he should look at each item on the list and ask the following questions:

- Will equipment be needed? Is it available?
- Will supervision be needed? Is it available?
- Are there safety factors to consider?

Answering these questions helps the designer decide which exercises to use. He can choose from the exercises, calisthenics, conditioning drills, grass drills, and guerrilla drills described in this chapter. However, he should not limit the circuit to only these activities. Imagination and field expediency are important elements in developing circuits that hold the interest of soldiers. (See Figures 7-1 through 7-3.)

Arrange the Stations

A circuit usually has 8 to 12 stations, but it may have as many as 20. After deciding how many stations to include, the designer must decide how to arrange them. For example, in a circuit for strength training, the same muscle group should not be exercised at consecutive stations.

One approach is to alternate "pushing" exercises with "pulling" exercises which involve movement at the same joint(s). For example, in a strength training circuit, exercisers may follow the pushing motion of a bench press with the pulling motion of the seated row. This could be followed by the pushing motion of the overhead press which could be followed by the pulling motion of the lat pull-down. Another approach might be to alternate between upper and lower body exercises.

The choice of exercises for circuit training depends on the objectives of the circuit.

By not exercising the same muscle group twice in a row, each muscle has a chance to recover before it is used in another exercise. If some exercises are harder than others, soldiers can alternate hard exercises with easier ones. The choice of exercises depends on the objectives of the circuit.

Select the Training Sites

Circuits may be conducted outdoors or indoors. If the designer wants to include running or jogging a certain distance between stations, he may do this in several ways. In the gymnasium, soldiers may run five laps or for 20 to 40 seconds between stations. Outdoors, they may run laps or run between spread-out stations if space is available. However, spreading the stations too far apart may cause problems with control and supervision.

Prepare a Sketch

The designer should draw a simple sketch that shows the location of each station in the training area. The sketch should include the activity and length of time at each station, the number of stations, and all other useful information.

Lay Out the Stations

The final step is to lay out the stations which should be numbered and clearly marked by signs or cards. In some cases, instructions for the stations are written on the signs. The necessary equipment is placed at each station.

Sample Conditioning Circuits

Figures 7-1, 7-2, and 7-3 show different types of conditioning circuits. Soldiers should work at each station 45 seconds and have 15 seconds to rotate to the next station.

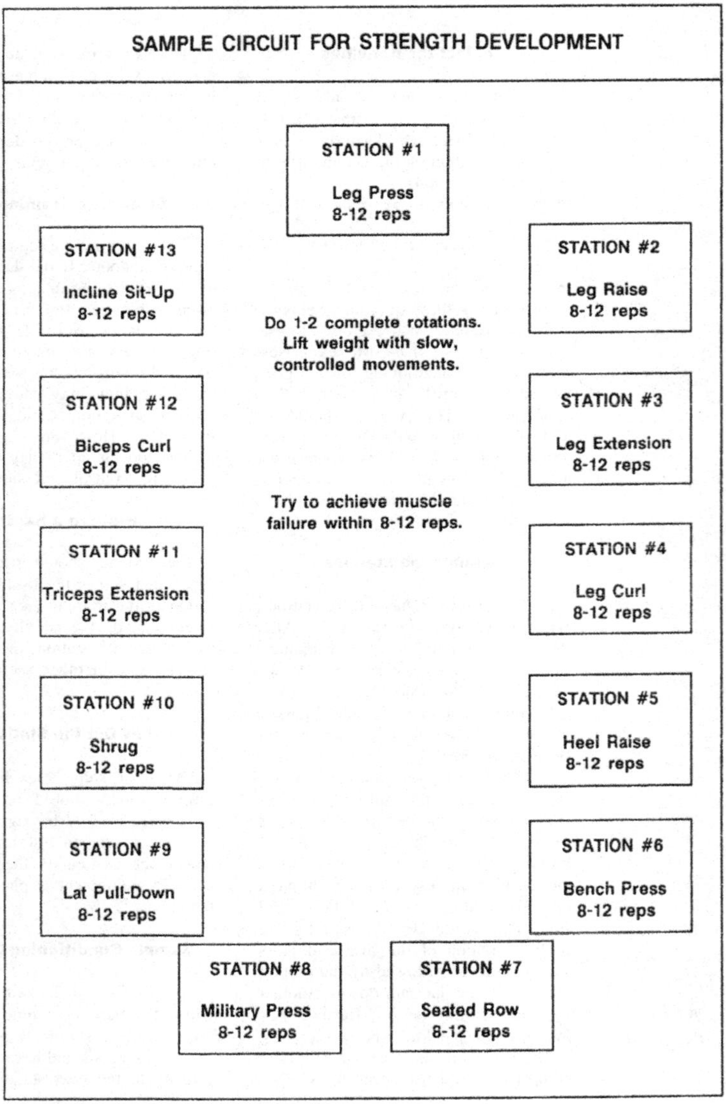
SAMPLE CIRCUIT FOR STRENGTH DEVELOPMENT
STATION #1
Leg Press
8-12 reps
STATION #2
Leg Raise
8-12 reps
STATION #3
Leg Extension
8-12 reps
STATION #4
Leg Curl
8-12 reps
STATION #5
Heel Raise
8-12 reps
STATION #6
Bench Press
8-12 reps
STATION #7
Seated Row
8-12 reps
STATION #8
Military Press
8-12 reps
STATION #9
Lat Pull-Down
8-12 reps
STATION #10
Shrug
8-12 reps
STATION #11
Triceps Extension
8-12 reps
STATION #12
Biceps Curl
8-12 reps
STATION #13
Incline Sit-Up
8-12 reps
Do 1-2 complete rotations.
Lift weight with slow,
controlled movements.
Try to achieve muscle
failure within 8-12 reps.

Figure 7-1

SAMPLE CIRCUIT FOR CARDIORESPIRATORY ENDURANCE

Do 2-3 complete rotations.

Stations may be 25-30 meters apart to allow more running.

STATION #1
Stationary Run
30 seconds

STATION #2
Push-Up
30 seconds

STATION #3
Side-Straddle Hop
30 seconds

STATION #4
Sit-Up
30 seconds

STATION #5
Ski Jumps
30 seconds

STATION #6
Flutter Kicks
30 seconds

STATION #7
Bend and Reach
(done slowly)
30 seconds

STATION #8
Wide-Hand Push-Ups
30 seconds

STATION #9
Bicycle
30 seconds

STATION #10
Knee Bender
30 seconds

STATION #11
Steam Engine
30 seconds

STATION #12
Twisting Sit-up
30 seconds

STATION #13
Mule Kicks
30 seconds

STATION #14
All-Fours Run
30 seconds

Figure 7-2

SAMPLE CIRCUIT FOR PUSH-UP AND SIT-UP IMPROVEMENT

STATION #1
Elevated Push-Up
30 seconds

STATION #2
Twisting Sit-Up
30 seconds

STATION #3
Parallel Dips
30 seconds

STATION #4
Sit-Up
30 seconds

STATION #5
Wide-Hand Push-Up
30 seconds

STATION #6
Flutter Kick
30 seconds

STATION #7
Close-Hand Push-Up
30 seconds

STATION #8
Bicycle
30 seconds

Do 1-2 complete rotations.

Time may decrease to 20 sec on the second rotation.

Move immediately from station to station. If too fatigued, push-ups may be done on the knees.

Figure 7-3

Calisthenics

Calisthenics can be used to help develop coordination. CR and muscular encurance, flexibility, and strength.

Calisthenics can be used to exercise most of the major muscle groups of the body. They can help develop coordination, CR and muscular endurance, flexibility, and strength. Poorly-coordinated soldiers, however, will derive the greatest benefit from many of these exercises

Although calisthenics have some value when included in a CR circuit or when exercising to music, for the average soldier, calisthenics such as the bend and reach, squat bender, lunger, knee bender, and side-straddle hop can best be used in the warm-up and cool-down periods. Exercises such as the push-up, sit-up, parallel bar dip, and chin-up/pull-up, on the other hand, can effectively be used in the conditioning period to develop muscular endurance or muscular strength.

Please note that exercises such as the bend and reach, lunger, and leg spreader, which were once deleted from FM 21-20 because of their potential risk to the exerciser, have been modified and reintroduced in this edition. All modifications should be strictly adhered to.

Few exercises are inherently unsafe. Nonetheless, some people, because of predisposing conditions or injuries, may find certain exercises less safe than others. Leaders must consider each of their soldier's physical limitations and use good judgment before letting a soldier perform these exercises. However, for the average soldier who is of sound body, following the directions written below will produce satisfactory results with a minimum risk of injury.

Finally, some of the calisthenics listed below may be done in cadence. These calisthenics are noted, and directions are provided below with respect to the actions and cadence. When doing exercises at a moderate cadence, use 80 counts per minute. With a slow cadence, use 50 counts per minute unless otherwise directed.

SAFETY FACTORS

While injury is always possible in any vigorous physical activity, few calisthenic exercises are really unsafe or dangerous. The keys to avoiding injury while gaining training benefits are using correct form and intensity. Also, soldiers with low fitness levels, such as trainees, should not do the advanced exercises highly fit soldiers can do. For example, with the lower back properly supported, flutter kicks are an excellent way to condition the hip flexor muscles. However, without support, the possibility of straining the lower back increases. It is not sensible to have recruits do multiple sets of flutter kicks because they probably are not conditioned for them. On the other hand, a conditioned Ranger company may use multiple sets of flutter' kicks with good results.

The key to doing calisthenic exercises safely is to use common sense. Also, ballistic (that is, quick-moving) exercises that combine rotation and bending of the spine increase the risk of back injury and should be avoided. This is especially true if someone has had a previous injury to the back. If this type of action is performed, slow stretching exercises, not conditioning drills done to cadence, should be used.

Some soldiers complain of shoulder problems resulting from rope climbing, horizontal ladder, wheelbarrow, and crab-walk exercises. These exercises are beneficial when the soldier is fit and he does them in a regular, progressive manner. However, a certain level of muscular strength is needed to do them safely. Therefore, soldiers should progressively train to build up to these exercises. Using such exercises for unconditioned soldiers increases the risk of injury and accident.

Progression and Recovery

Other important principles for avoiding injury are progression and recovery. Programs that try to do too much too soon invite problems. The day after a "hard" training day, if soldiers are working the same muscle groups and/or fitness components, they should work them at a reduced intensity to minimize stress and permit recovery.

The best technique is to train alternate muscle groups and/or fitness components on different days. For example, if the Monday-Wednesday-Friday (M-W-F) training objective is CR fitness, soldiers can do ability group running at THR with some light calisthenics and stretching. If the Tuesday-Thursday (T-Th) objective is muscular endurance and strength, soldiers can benefit from doing partner-resisted exercises followed by a slow run. To ensure balance and regularity in the program, the next week should have muscle endurance and strength development on M-W-F and training for CR endurance on T-Th. Such a program has variety, develops all the fitness components, and follows the seven principles of exercise while, at the same time, it minimizes injuries caused by overuse.

Leaders should plan PT sessions to get a positive training effect, not to conduct "gut checks." They should know how to correctly do all the exercises in their program and teach their soldiers to train using good form to help avoid injuries.

Key Points for Safety

Doing safe exercises correctly improves a soldier's fitness with a minimum risk of injury.

The following are key points for ensuring safety during stretching and calisthenic exercises:

- Stretch slowly and without pain and unnatural stress to a joint. Use static (slow and sustained) stretching for warming up, cooling down, ballistic (bouncy or jerky) stretching movements.
- Do not allow the angle formed by the upper and lower legs to become less than 90 degrees when the legs are bearing weight.
- A combination of spinal rotation and bending should generally be avoided. However, if done, use only slow, controlled movements with little or no extra weight.

Leaders must be aware of the variety of methods they may use to attain their physical training goals. The unit's Master Fitness Trainer is schooled to provide safe, effective training methods and answer questions about training techniques.

CALISTHENIC EXERCISES

The following are some common calisthenic exercises.

SIDE-STRADDLE HOP

Position: Assume the position of attention.
Action: (1) Jump slightly into the air while moving the legs more than shoulder-width apart, swinging the arms overhead, and clapping the palms together. (2) Jump slightly into the air while swinging the arms sideward and downward and returning to the position of attention. (3) Repeat action 1. (4) Repeat action 2. Use a moderate cadence.

Variation: (1) Jump slightly into the air while moving the left leg forward and the right leg backward, swinging the arms overhead, and clapping the palms together. (2) Jump slightly into the air while swinging the arms sideward and downward and returning to the position of attention. (3) Repeat the jumping and arm movements of action 1 while moving the right leg forward and the left leg backward. (4) Repeat action 2. Use a moderate cadence.

MULE KICK

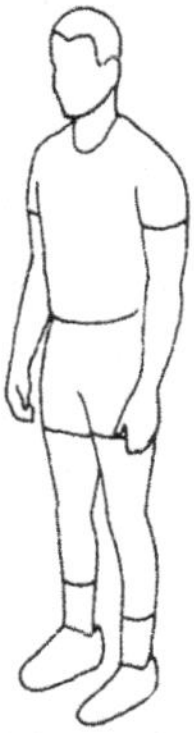

Position: Stand with the feet shoulder-width apart.

Action: Jump up repeatedly while kicking the heels to the buttocks. To do the Mule Kick to cadence, do one repetition per count. Use a moderate cadence.

SKI JUMP

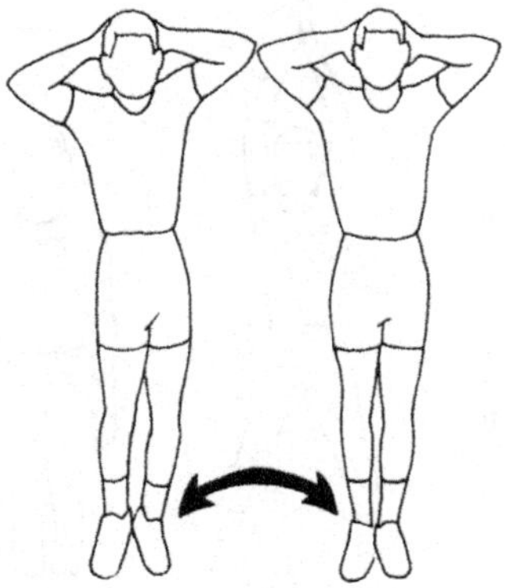

Position: Stand with the feet together, the hands placed behind the head with the fingers interlaced.

Action: (1) Keeping the feet together, jump sideways to the left. (2) Keeping the feet together, jump sideways to the right. (3) Repeat action 1. (4) Repeat action 2. Use a moderate cadence.

FLUTTER KICK

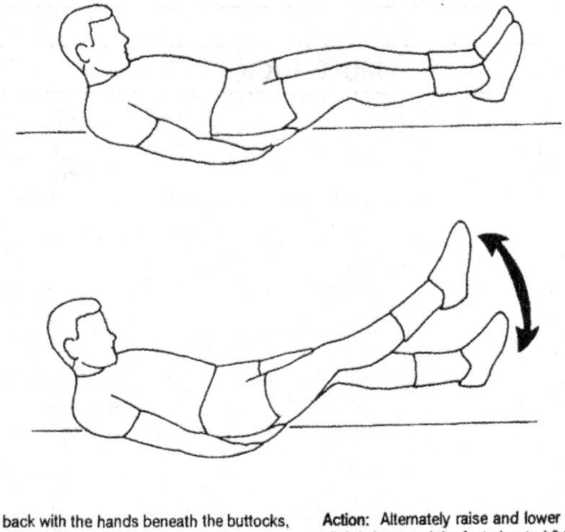

Position: Lie on your back with the hands beneath the buttocks, the head raised, and the knees slightly bent.

Action: Alternately raise and lower the legs, keeping the knees slightly bent and the feet elevated 6 to 18 inches above the floor. To do the flutter kick to cadence, do one repetition per count. Use a moderate cadence.

BEND AND REACH

Position: Stand in a wide, side-straddle position with the palms facing each other and the arms overhead and straight.

Action: (1) Bend at the knees and waist. **Slowly** bring the arms down, and reach between the legs as far as possible. **Make sure the angle formed by the upper and lower leg is never less than 90 degrees.** (2) Recover slowly to the start position. (3) Repeat action 1. (4) Repeat action 2. Use a slow cadence.

HIGH JUMPER

Position: Place the feet about shoulder-width apart with the knees flexed. Bend forward at the waist, aligning the arms with the trunk and hips. Keep the arms straight at all times during the exercise. Keep the palms facing each other with the head and eyes initially to the front.

Action: (1) Take a slight jump into the air while swinging the arms forward and up to shoulder level. (2) Take a slight jump while swinging the arms backward, returning to the start position. (3) Jump strongly upward while swinging the arms forward and up to the overhead position; at the same time, briefly look skyward. While descending, return the head and eyes to the front, and flex the knees. (4) Repeat action 2. Use a moderate cadence.

SQUAT BENDER

Position: Stand with the feet shoulder-width apart, hands on hips, thumbs in the small of the back, and the elbows back.

Action: (1) Bending the knees, lower yourself to a half-squat position while maintaining balance on the balls of the feet. With the trunk inclined slightly forward, thrust the arms forward to shoulder level with the elbows locked and the palms down. (2) Recover to the start position. (3) Keeping the knees slightly bent, bend forward at the waist, touching the ground in front of the toes. (4) Recover to the start position. Use a moderate cadence.

LUNGER

Position: Start from the position of attention.
Action: (1) Lunge diagonally forward to the left by stepping in that direction with the left foot, placing the left knee over the left foot. At the same time, place the arms sideward at shoulder level, the palms up, and the head and shoulders squarely to the front.

(2) Bend slowly forward and downward over the left thigh, and wrap the arms around the thigh, hands grasping the opposite arms above the elbows. (3) Recover slowly to the second position by releasing the arms, straightening the trunk, and extending the arms sideward, palms up. (4) Resume the position of attention by dropping the arms and returning the left foot to the side of the right. Repeat the exercise to the right side. Use a moderate cadence.

KNEE BENDER

Position: Stand with the feet shoulder-width apart, hands on the hips, the thumbs in the small of the back, and the elbows back.

Action: Bend at the knees, lean slightly forward at the waist with the head up, and slide the hands along the outside of the legs until the extended fingers reach the top of the boots or the middle of the lower leg. (2) Recover to the start position. (3) Repeat action 1. (4) Repeat action 2. Use a moderate cadence.

THE SWIMMER

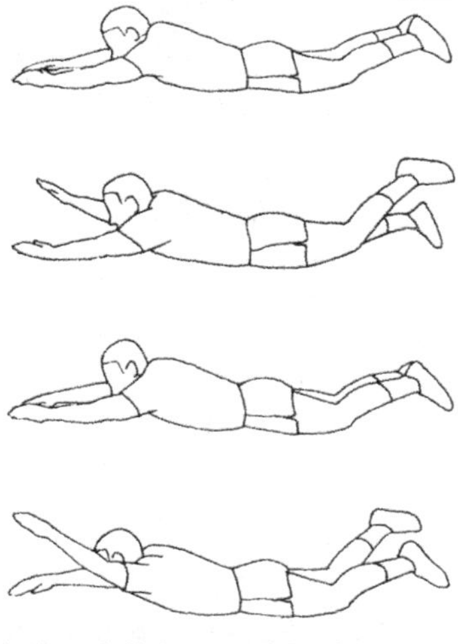

Position: Lie prone with the feet together and with the arms together and extended forward in front of the body. Keep the arms and legs straight at all times during this exercise.

Action: (1) Move the right arm and left leg up. (2) Return to the start position. (3) Move the left arm and right leg up. (4) Return to the start position. Continue in an alternating manner. Use a moderate cadence.

SUPINE BICYCLE

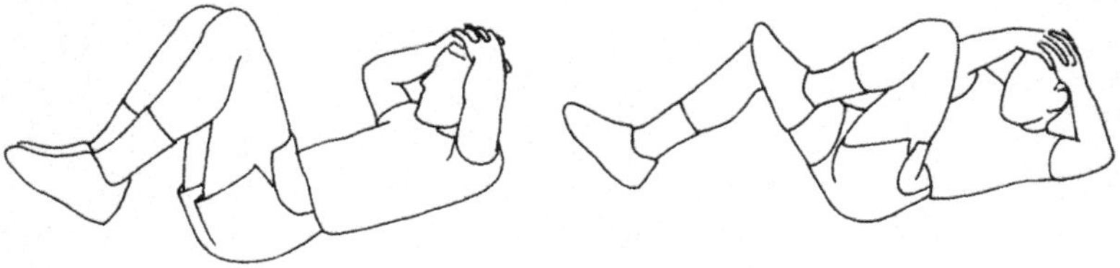

Position: Assume a supine position with the hips and knees flexed. Place the palms directly on top of the head with the fingers interlaced.

Action: (1) Bring the left knee upward while curling the trunk upward, and touch the right elbow to the left knee. (2) Repeat action 1 with the other leg and elbow. (3) Repeat action 1. (4) Repeat action 2. Use a slow cadence.

THE ENGINE

Position: Stand with the arms straight and in front of the body. The arms should be parallel to the ground with the palms facing downward.

Action: (1) Bring the left knee upward to the left elbow. (2) Return to the start position. (3) Touch the right knee to the right elbow. (4) Recover to the start position. Be sure to keep the arms parallel to the ground throughout the entire exercise. Use a moderate cadence.

CROSS-COUNTRY SKIER

Position: Assume a position of attention.
Action: Jump slightly into the air, and move the left foot forward and the right foot backward, landing with both knees slightly bent. At the same time, move the right arm upward and forward to shoulder height and the left arm back as far as possible, always keeping the arms straight and the palms facing each other. (2) Jump slightly into the air, and move the right foot forward and the left foot backward. At the same time, move the left arm upward and forward to shoulder height and the right arm back as far as possible. (3) Repeat action 1. (4) Repeat action 2. Use a moderate cadence.

PUSH-UP

Position: Assume the front-leaning rest position with the hands placed comfortably apart, the feet together or up to 12 inches apart, and the body forming a generally straight line from the shoulders to the ankles.
Action: Keeping the body straight throughout the exercise, lower the body until the upper arms are at least parallel to the ground. Then, push yourself up to the initial position by completely straightening the arms.

Push-Up Variations: To train the muscles more completely, place the hands at varying widths. They may be wider apart or closer together than shoulder width. Elevating the feet to different heights makes push-ups more difficult. The higher the feet, the more difficult the exercise. Push-ups are also more difficult when the hands and feet are placed on boxes or chairs. This helps the soldier exercise through a fuller range of motion. To do extra repetitions when fatigued, drop to the knees while keeping the knees, hips, and shoulders in a straight line.

SIT-UP

Position: Lie on the back with the feet together or up to 12 inches apart, the knees bent so that an angle of 90 degrees is formed by the upper and lower legs, and the fingers interlocked behind the head.
Action: Raise your upper body forward to the vertical position so that the base of the neck is above the base of the spine, then lower yourself in a controlled manner until the bottom of the shoulder blades touch the ground.

Sit-Up Variations: Variations include keeping the feet elevated and crossing the hands on the chest.

CHIN-UP (PULL-UP)

Position: Hang from a horizontal bar with the arms fully extended.
Action: Bend your elbows and pull yourself upward until your chin is above the bar; do not swing or kick your legs. Return to the starting position in a controlled manner.

Variations: Use overhand (pull-up), underhand (chin-up), or alternating grips, with the hands close together, far apart, or at shoulder-width. If unable to complete a chin-up using proper form, elevate yourself to the up position with help and hang there, or slowly lower yourself to the starting position. Repeat this several times, gradually adding more repetitions from workout to workout.

PARALLEL BAR DIP

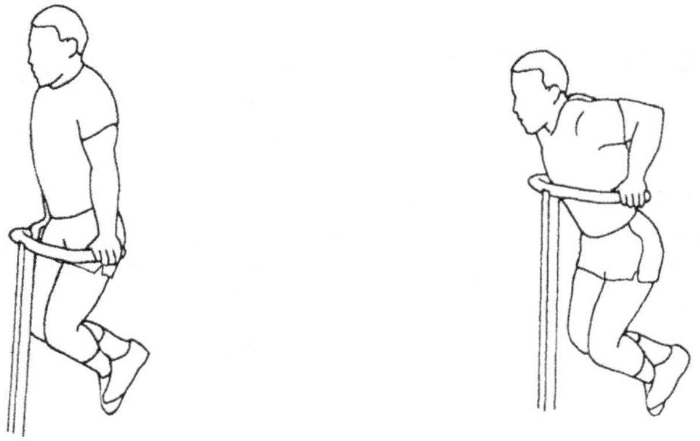

Position: Keep the feet off the floor and support the body's weight on straight arms.

Action: Bend the arms and lower the body in a controlled manner until the upper arms are at least parallel to the floor. If necessary, bend the legs at the knees to keep the feet from touching the floor. Straighten the arms to return to the starting position.

CONDITIONING DRILLS

Conditioning drills are intended to supplement muscular strength and endurance training sessions.

Some large units prefer to use sets of calisthenic exercises as part of their PT sessions. Figure 7-4 shows three calisthenic conditioning drills for both the poorly conditioned and physically fit soldiers. The drills are designed to be done progressively and are intended to supplement muscular strength and endurance training sessions.

Leaders can mix the exercises to provide greater intensity, based on the fitness level of the soldiers being trained. However, they should choose and sequence them to alternate the muscle groups being worked. Soldiers should do each exercise progressively from 15 to 40 or more repetitions (20 to 60 seconds for timed sets) based on their level of conditioning. They may also do each exercise in cadence unless timed sets are specified. For timed sets, soldiers do as many repetitions of an exercise as possible in the allowed time. Using timed sets, both the well-conditioned and less-fit soldiers can work themselves to their limits.

The following conditioning drills (Figure 7-4) are arranged according to the phase of training.

Grass Drills

Grass drills are exercise movements that feature rapid changes in body position. These are vigorous drills which, when properly done, exercise all the major muscle groups. Soldiers should respond to commands as fast as possible and do all movements at top speed. They continue to do multiple repetitions of each exercise until the next command is given. No cadence is counted.

TRAINING-PHASE CONDITIONING DRILLS

#1 PREPARATORY TRAINING

High Jumper
Push-Up (TS* 20-45** seconds)
Sit-Up (TS 20-45** seconds)
Side-Straddle Hop
Side Bender
Knee Bender
Stationary Run

#2 CONDITIONING TRAINING

Push-UP (varied hand positions)
 (TS 30-60 seconds)
Supine Bicycle
High Jumper
Sit-Up (all types)
 (TS 30-60 seconds)
The Engine or Cross-Country Skiier
All-Fours Run (stationary)

#3 MAINTENANCE TRAINING

Ski Jump
Sit-Ups (all types) (TS 30-60 seconds)
Push-Up (varied hand positions) (TS 30-60 seconds)
Mule Kick
Flutter Kick
The Engine
The Swimmer

*TS = timed set

** Because of a lower level of fitness, 45 seconds will usually be the upper limit.

Figure 7-4

Performing grass drills can improve CR endurance, help develop muscular endurance and strength, and speed up reaction time. Since these drills are extremely strenuous, they should last for short periods (30 to 45 seconds per exercise). The two drills described here each have four exercises. Leaders can develop additional drills locally.

The soldiers should do a warm-up before performing the drills and do a cool-down afterward. The instructor does all the activities so that he can gauge the intensity of the session. The commands for grass drills are given in rapid succession without the usual preparatory commands. To prevent confusion, commands are given sharply to distinguish them from comments or words of encouragement.

As soon as the soldiers are familiar with the drill, they do all the exercises as vigorously and rapidly as possible, and they do each exercise until the

Grass drills are exercise movements that feature rapid changes in body position.

Soldiers should do a warm-up before performing grass drills and do a cool-down afterward.

Progression with grass drills is made by a gradual increase in the time devoted to the drills.

next command is given. Anything less than a top-speed performance decreases the effectiveness of the drills.

Once the drills start, soldiers do not have to resume the position of attention. The instructor uses the command "Up" to halt the drill for instructions or rest. At this command, soldiers assume a relaxed, standing position.

Grass drills can be done in a short time. For example, they may be used when only a few minutes are available for exercise or when combined with another activity. Sometimes, if time is limited, they are a good substitute for running.

Most movements are done in place. The extended-rectangular formation is best for a platoon- or company-sized unit. The circle formation is more suitable for squad- or section-sized groups.

When soldiers are starting an exercise program, a 10- to 15-minute workout may be appropriate. Progression is made by a gradual increase in the time devoted to the drills. As the fitness of the soldiers improves, the times should be gradually lengthened to 20 minutes. The second drill is harder than the first. Therefore, as soldiers progress in the first drill, the instructor should introduce the second. If he sees that the drill needs to be longer, he can repeat the exercises or combine the two drills.

STARTING POSITIONS

After the warm-up, bring the soldiers to a position of ATTENTION. The drills begin with the command *GO.* Other basic commands are FRONT, BACK, and STOP. (See Figure 7-5 for the positions and actions associated with these commands.)

- ATTENTION: The position of at tention is described in FM 22-5, Drill and Ceremonies.
- GO This involves running in place at top speed on the balls of the feet. The soldier raises his knees high, pumps his arms, and bends forward slightly at the waist.
- FRONT The soldier lies prone with elbows bent and palms directly under the shoulders as in the down position of the push up. The legs are straight and together with the head toward the instructor.
  BACK: The soldier lies flat on his back with his arms extended along his sides and his palms facing down ward. His legs are straight and to gether; his feet face the instructor.
- STOP The soldier assumes the stance of a football lineman with feet spread and staggered. His left arm is across his left thigh; his right arm is straight. His knuckles are on the ground; his head is up, and his back is roughly parallel to the ground.

To assume the FRONT or BACK position from the standing GO or STOP positions, the soldier changes positions vigorously and rapidly. (See Figure 7-5.)

To change from the FRONT to the BACK position (Figure 7-5), the soldier does the following:

- Takes several short steps to the right or left.
  Lifts his arm on the side toward which his feet move.
- Thrusts his legs vigorously to the front.

To change from the BACK to the FRONT position, the soldier sits up quickly. He places both hands on the ground to the right or left of his legs. He takes several short steps to the rear on the side opposite his hands. When his feet are opposite his hands, he thrusts his legs vigorously to the rear and lowers his body to the ground. (See Figure 7-5.)

STARTING POSITIONS FOR GRASS DRILLS

Figure 7-5

GRASS DRILL ONE

Exercises for grass drill one are described below and shown in Figure 7-6.

Bouncing Ball

From the FRONT position, push up and support the body on the hands (shoulder-width apart) and feet. Keep the back and legs generally in line and the knees straight. Bounce up and down in a series of short, simultaneous, upward springs from the hands, hips, and feet.

Supine Bicycle

From the BACK position, flex the hips and knees. Place the palms directly on top of the head, and interlace the fingers. Bring the knee of one leg upward toward the chest. At the same time, curl the trunk and head upward while touching the opposite elbow to the elevated knee. Repeat with the other leg and elbow. Continue these movements as opposite legs and arms take turns.

Knee Bender

From the position of ATTENTION, do half-knee bends with the feet in line and the hands at the sides. Make sure the knees do not bend to an angle less than 90 degrees.

Roll Left and Right

From the FRONT position, continue to roll in the direction commanded until another command is given. Then, return to the FRONT position.

GRASS DRILL TWO

Exercises for grass drill two are described below and shown in Figure 7-6.

The Swimmer

From the FRONT position, extend the arms forward. Move the right arm and left leg up and down; then, move the left arm and right leg up and down. Continue in an alternating manner.

Bounce and Clap Hands

The procedure is almost the same as for the bouncing ball in grass drill one. However, while in the air, clap the hands. This action requires a more vigorous bounce or spring. The push-up may be substituted for this exercise.

Leg Spreader

From the BACK position, raise the legs until the heels are no higher than six inches off the ground. Spread the legs apart as far as possible, then put them back together. Keep the head off the ground. Throughout, place the hands under the upper part of the buttocks, and slightly bend the knees to ease pressure on the lower back. Open and close the legs as fast as possible. The curl-up may be substituted for this exercise.

Forward Roll

From the STOP position, place both hands on the ground, tuck the head, and roll forward. Keep the head tucked while rolling.

Stationary Run

From the position of ATTENTION, start running in place at the GO command by lifting the left foot first. Follow the instructor as he counts two repetitions of cadence. For example, "One, two, three, four; one, two, three, four." The instructor then gives informal commands such as the following: "Follow me," "Run on the toes and balls of your feet," "Speed it up," "Increase to a sprint, raise your knees high, lean forward at your waist, and pump your arms vigorously," and "Slow it down."

To halt the exercise, the instructor counts two repetitions of cadence as the left foot strikes the ground: "One, two, three, four, one, two, three, HALT."

Figure 7-6

Guerilla Exercises

Soldiers progress with guerilla exercises by shortening the quick-time marching periods between exercises and by doing all the exercises a second time.

Guerrilla exercises, which can be used to improve agility, CR endurance, muscular endurance, and to some degree muscular strength, combine individual and partner exercises. These drills require soldiers to change their positions quickly and do various basic skills while moving forward. Figures 7-7 and 7-8 show these exercises.

The instructor decides the duration for each exercise by observing its effect on the soldiers. Depending on how vigorously it is done, each exercise should be continued for 20 to 40 seconds.

The group moves in circle formation while doing the exercises. If the platoon exceeds 30 soldiers, concentric circles may be used. A warm-up activity should precede these exercises, and a cool-down should follow them. After the circle is formed, the instructor steps into the center and issues commands.

EXERCISE AND PROGRESSION

Soldiers progress by shortening the quick-time marching periods between exercises and by doing all exercises a second time. This produces an *overload* that improves fitness.

Many soldiers have not had a chance **to** do the simple skills involved in guerrilla exercises. However, they can do these exercises easily and quickly in almost any situation.

The preparatory command is always the name of the exercise, and the command of execution is always "March." The command "Quick time, march" ends each exercise.

For the double guerrilla exercises (in circle formation) involving two soldiers, the commands for pairing are as follows:

Figure 7-7

- "Platoon halt."
- "From (soldier is designated), by twos, count off." (For example: 1-2, 1-2, 1-2.)
- "Even numbers, move up behind odd numbers." (Pairs are adjusted according to height and weight.)
- "You are now paired up for double guerrillas." The command "Change" is given to change the soldiers' positions.

After the exercises are completed, the instructor halts the soldiers and positions the base soldier or platoon guide by commanding, "Base man (or platoon guide), post." He then commands "Fall out and fall in on the base man (or platoon guide)."

EXERCISE DESCRIPTIONS

Brief explanations of guerrilla exercises follow.

All-Fours Run

Face downward, supporting the body on the hands and feet. Advance forward as fast as possible by moving the arms and legs forward in a coordinated way.

Bottoms-Up Walk

Take the front-leaning rest position, and move the feet toward the hands in short steps while keeping the knees locked. When the feet are as close to the hands as possible, walk forward on the hands to the front-leaning-rest position.

Crab Walk

Assume a sitting position with the hips off the ground and hands and feet supporting the body's weight. Walk forward, feet first.

The Engine

Stand with the arms straight and in front of the body. The arms should be parallel to the ground with the palms facing downward. While walking forward, bring the left knee upward to the left elbow. Return to the start position. Continuing to walk forward, touch the right knee to the right elbow. Recover to the start position. Be sure to keep the arms parallel to the ground throughout the entire exercise.

Double Time

Do a double-time run while maintaining the circle formation.

Broad Jump

Jump forward on both feet in a series of broad jumps. Swing the arms vigorously to help with the jumps.

Straddle Run

Run forward, leaping to the right with the left foot and to the left with the right foot.

Hobble Hopping

Hold one foot behind the back with the opposite hand and hop forward. On the command "Change," grasp the opposite foot with the opposite hand and hop forward.

Two-Man Carry

For two-man carries, soldiers are designated as number one (odd-numbered) and number two (even-numbered). A number-one and number-two soldier work as partners.

Fireman's Carry

Two soldiers do the carry. On command, number-two soldier bends at the waist, with feet apart in a balanced stance. Number-one soldier moves toward his partner. He places himself by his partner's left shoulder and bends himself over his partner's shoulders and back. When in position, number-two soldier, with his left hand, reaches between his partner's legs and grasps his left wrist. On command, they move forward until the command for changeover. They then change positions. The fireman's carry can also be done from the other side.

Single-Shoulder Carry

Two soldiers do the carry. On command, number-two soldier bends at the waist with feet apart in a balanced stance. At the same time, number-one soldier moves toward his partner. He places his abdominal area onto his partner's right or left shoulder and leans over. Number-two soldier puts his arms around the back of his partner's knees and stands up. On command, they move forward until the command for changeover. They then change positions.

Cross Carry

On command, number-two soldier bends over at the waist. He twists slightly to the left with feet spread apart in a balanced position. At the same time, number-one soldier moves toward his partner's left side and leans over his partner's back. Number two soldier, with his left arm, reaches around his partner's legs. At the same time, he reaches around his partner's back with his right arm, being careful not to grab his partner's neck or head. He then stands up straight, holding his partner on his back. On command, they move forward until the command for changeover. They then change positions.

Saddle-Back (Piggyback) Carry

On command, number-two soldier bends at the waist and knees with his hand on his knees and his head up. To assume the piggyback position, number-one soldier moves behind his partner, places his hands on his partner's shoulders, and climbs carefully onto his partner's hips. As number-one soldier climbs on, number-two soldier grasps his partner's legs to help support him. Number-one soldier places his arms over his partner's shoulders and crosses his hands over his partner's upper chest. They move forward until the command for changeover is given. They then change positions.

ADDITIONAL GUERILLA EXERCISES
DOUBLE TIME
BROAD JUMP
STRADDLE RUN
HOBBLE HOPPING
FIREMAN'S CARRY
SINGLE-SHOULDER CARRY
CROSS CARRY
SADDLE-BACK CARRY

Figure 7-8

CHAPTER 8

Obstacle Courses and Additional Drills

This chapter describes obstacle courses as well as rifle drills, log drills, and aquatic exercises. These are not designed to develop specific components of physical fitness. Commanders should use them to add variety to their PT programs and to help soldiers develop motor fitness including speed, agility, coordination, and related skills and abilities. Many of these activities also give soldiers the chance to plan strategy, make split-second decisions, learn teamwork, and demonstrate leadership.

Obstacle Courses

There are two types of obstacle courses-conditioning and confidence.

Physical performance and success in combat may depend on a soldier's ability to perform skills like those required on the obstacle course. For this reason, and because they help develop and test basic motor skills, obstacle courses are valuable for physical training.

There are two types of obstacle courses--conditioning and confidence. The conditioning course has low obstacles that must be negotiated quickly. Running the course can be a test of the soldier's basic motor skills and physical condition. After soldiers *receive* instruction and practice the skills, they run the course against time.

A confidence course has higher, more difficult obstacles than a conditioning course. It gives soldiers confidence in their mental and physical abilities and cultivates their spirit of daring. Soldiers are encouraged, but not forced, to go through it. Unlike conditioning courses, confidence courses are not run against time.

NONSTANDARD COURSES AND OBSTACLES

Commanders may build obstacles and courses that are nonstandard (that is, not covered in this manual) in order to create training situations based on their unit's METL. When planning and building such facilities, designers should, at a minimum, consider the following guidance:

- Secure approval from the local installation's commander.
- Prepare a safety and health-risk assessment to support construction of each obstacle.
- Coordinate approval for each obstacle with the local or supporting safety office. Keep a copy of the approval in the permanent records.
- Monitor and analyze all injuries.
- Inspect all existing safety precautions on-site to verify their effectiveness.
- Review each obstacle to determine the need for renewing its approval.

SAFETY PRECAUTIONS

Instructors must always be alert to safety. They must take every precaution to minimize injuries as soldiers go through obstacle courses. Soldiers must do warm-up exercises before they begin. This prepares them for the physically demanding tasks ahead and helps minimize the chance of injury. A cool-down after the obstacle course is also necessary, as it helps the body recover from strenuous exercise.

Commanders should use ingenuity in building courses, making good use of streams, hills, trees, rocks, and other natural obstacles. They must inspect courses for badly built obstacles, protruding nails, rotten logs, unsafe landing pits, and other safety hazards.

There are steps which designers can take to reduce injuries. For example, at the approach to each obstacle, they should post an instruction board or sign with text and pictures showing how to negotiate it. Landing pits for jumps or vaults, and areas under or around obstacles where soldiers may fall from a height, should be filled with loose sand or sawdust, All landing areas should be raked and refilled before each use. Puddles of water under obstacles can cause a false sense of security. These could result in improper landing techniques and serious injuries. Leaders should postpone training on obstacle courses when wet weather makes them slippery.

Units should prepare their soldiers to negotiate obstacle courses by doing conditioning exercises beforehand. Soldiers should attain an adequate level of conditioning before they run the confidence course, Soldiers who have not practiced the basic skills or run the conditioning course should not be allowed to use the confidence course.

Instructors must explain and demonstrate the correct ways to negotiate all obstacles before allowing soldiers to run them. Assistant instructors should supervise the negotiation of higher, more dangerous obstacles. The emphasis is on avoiding injury. Soldiers should practice each obstacle until they are able to negotiate it. Before they run the course against time, they should make several slow runs while the instructor watches and makes needed corrections. Soldiers should never be allowed to run the course against time until they have practiced on all the obstacles.

CONDITIONING OBSTACLE COURSES

If possible, an obstacle course should be shaped like a horseshoe or figure eight so that the finish is close to the start. Also, signs should be placed to show the route.

A course usually ranges from 300 to 450 yards and has 15 to 25 obstacles that are 20 to 30 yards apart. The obstacles are arranged so that those which exercise the same groups of muscles are separated from one another.

The obstacles must be solidly built. Peeled logs that are six to eight inches wide are ideal for most of them. Sharp points and corners should be eliminated, and landing pits for jumps or vaults must be filled with sand or sawdust. Courses should be built and marked so that soldiers cannot sidestep obstacles or detour around them. Sometimes, however, courses can provide alternate obstacles that vary in difficulty.

Each course should be wide enough for six to eight soldiers to use at the same time, thus encouraging competition. The lanes for the first few obstacles should be wider and the obstacles easier than those that follow. In this way, congestion is avoided and soldiers can spread out on the course. To minimize the possibility of falls and injuries due to fatigue, the last two or three obstacles should not be too difficult or involve high climbing.

Trainers must always be aware that falls from the high obstacles could cause serious injury. Soldiers must be in proper physical condition, closely supervised, and adequately instructed.

The best way for the timer to time the runners is to stand at the finish and call out the minutes and seconds as each soldier finishes. If several watches are available, each wave of soldiers is timed separately. If only one watch is available, the waves are started at regular intervals such as every 30 seconds. If a soldier fails to negotiate an obstacle, a previously determined penalty is imposed.

When the course is run against time, stopwatches, pens, and a unit roster are needed. Soldiers may run the course with or without individual equipment.

Obstacles for Jumping

These obstacles are ditches to clear with one leap, trenches to jump into, heights to jump from, or hurdles. (See Figure 8-1.)

Instructors must explain and demonstrate the correct ways to negotiate all obstacles before allowing soldiers to run them.

Obstacles for Dodging

These obstacles are usually mazes of posts set in the ground at irregular intervals. (See Figure 8-2.) The spaces between the posts are narrow so that soldiers must pick their way carefully through and around them. Lane guides are built to guide soldiers in dodging and changing direction.

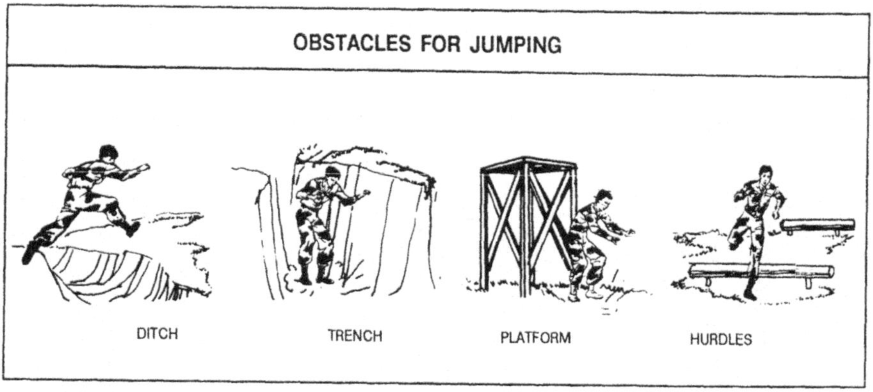

Figure 8-1

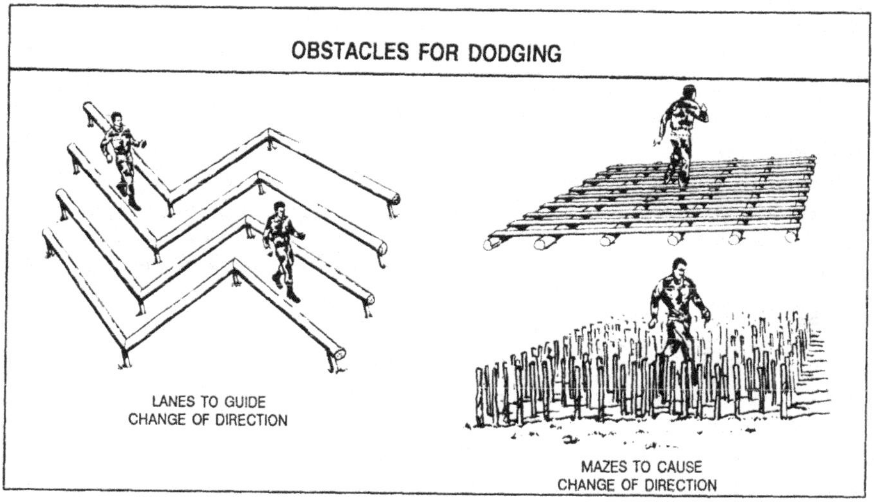

Figure 8-2

Obstacles for Vertical Climbing and Surmounting

These obstacles are shown at Figure 8-3 and include the following:

- Climbing ropes that are 1 1/2 inches wide and either straight or knotted.
- Cargo nets.
- Walls 7 or 8 feet high.
- Vertical poles 15 feet high and 6 to 8 inches wide.

Obstacles for Horizontal Traversing

Horizontal obstacles may be ropes, pipes, or beams. (See Figure 8-4.)

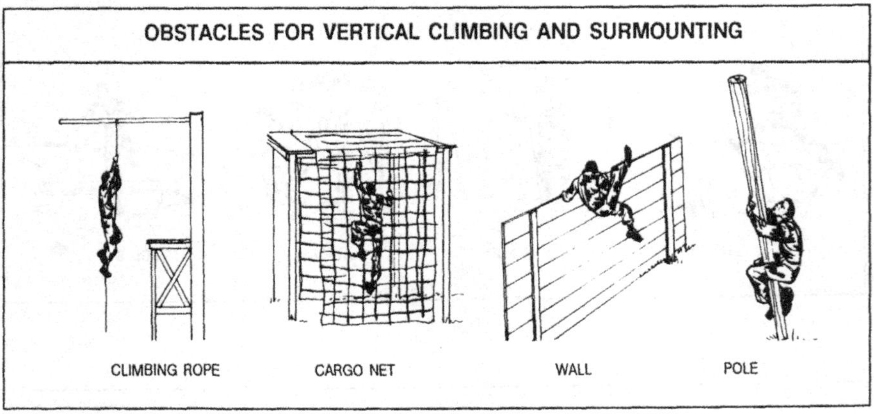

Figure 8-3

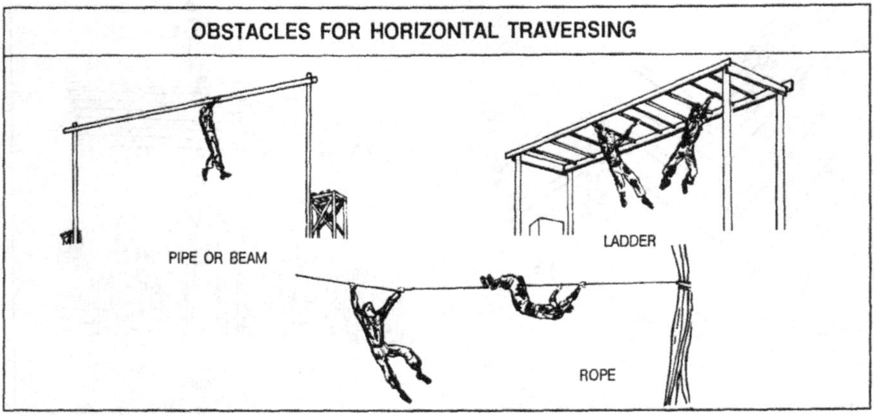

Figure 8-4

Obstacles for Crawling

These obstacles may be built of large pipe sections, low rails, or wire. (See Figure 8-5.)

Obstacles for Vaulting

These obstacles should be 3 to 3 1/2 feet high. Examples are fences and low walls. (See Figure 8-6.)

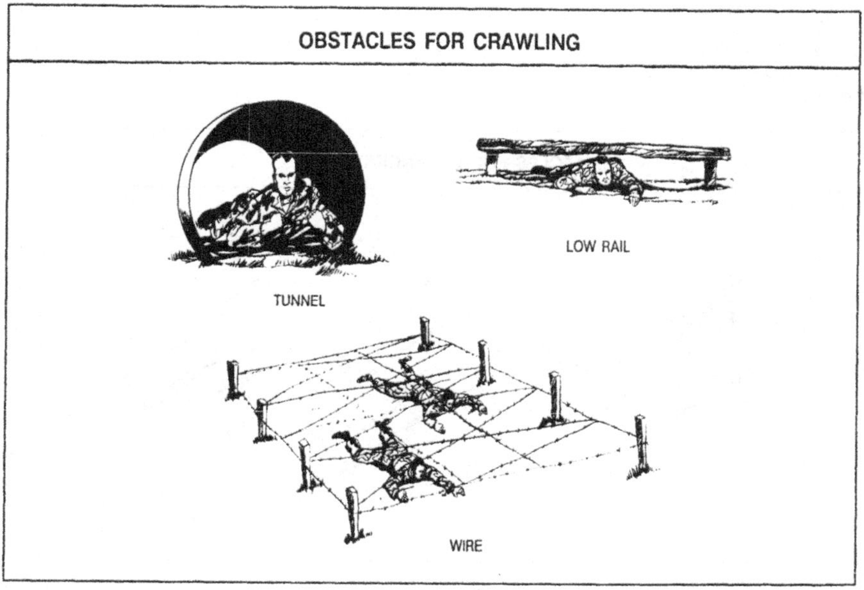

Figure 8-5

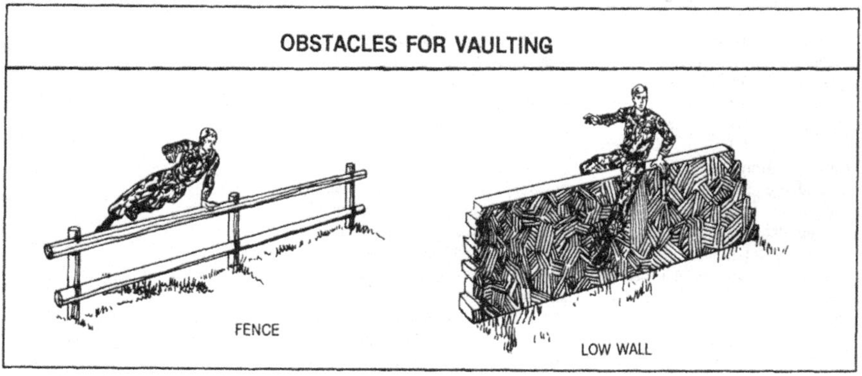

Figure 8-6

Obstacles for Balancing

Beams, logs, and planks may be used. These may span water obstacles and dry ditches, or they may be raised off the ground to simulate natural depressions. (See Figure 8-7.)

OBSTACLES FOR BALANCING

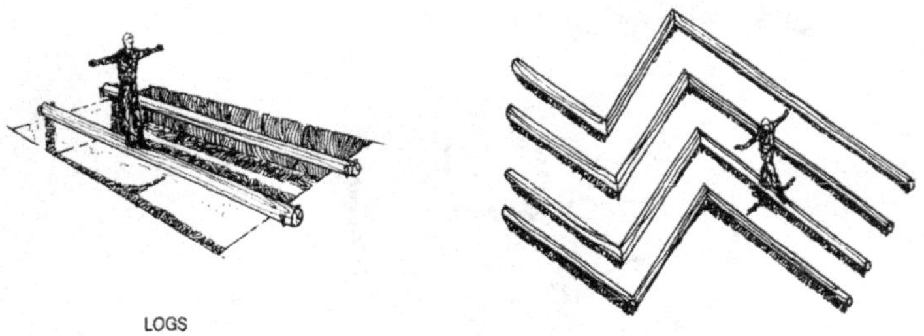

Figure 8-7

CONFIDENCE OBSTACLE COURSES

Confidence obstacle courses must be built in accordance with Folio No. 1, "Training Facilities," Corps of Engineers Drawing Number 28-13-95. You can obtain this publication from the Directorate of Facilities Engineering at most Army installations.

Confidence courses can develop confidence and strength by using obstacles that train and test balance and muscular strength. Soldiers do not negotiate these obstacles at high speed or against time. The obstacles vary from fairly easy to difficult, and some are high. For these, safety nets are provided. Soldiers progress through the course without individual equipment. Only one soldier at a time negotiates an obstacle unless it is designed for use by more than one.

Confidence courses should accommodate four platoons, one at each group of six obstacles. Each platoon begins at a different starting point. In the example below, colors are used to group the obstacles. Any similar method may be used to spread a group over the course. Soldiers are separated into groups of 8 to 12 at each obstacle. At the starting signal, they proceed through the course.

Soldiers may skip any obstacle they are unwilling to try. Instructors should encourage fearful soldiers to try the easier obstacles first. Gradually, as their confidence improves, they can take their places in the normal rotation. Soldiers proceed from one obstacle to the next until time is called. They then assemble and move to the next group of obstacles.

Rules for the Course

Supervisors should encourage, but not force, soldiers to try every obstacle. Soldiers who have not run the course before should receive a brief orientation at each obstacle, including an explanation and demonstration of the best way to negotiate it. Instructors should help those who have problems. Trainers and soldiers should not try to make obstacles more difficult by shaking ropes, rolling logs, and so forth. Close supervision and common sense must be constantly used to enhance safety and prevent injuries.

Soldiers need not conform to any one method of negotiating obstacles, but there is a uniformity in the general approach. Recommended ways to negotiate obstacles are described below.

Red Group

This group contains the first six obstacles. These are described below and numbered 1 through 6 in Figure 8-8.

Belly Buster. Soldiers vault, jump, or climb over the log. They must be warned that it is not stationary. Therefore, they should not roll or rock the log while others are negotiating it.

Reverse Climb. Soldiers climb the reverse incline and go down the other side to the ground.

Weaver. Soldiers move from one end of the obstacle to the other by weaving their bodies under one bar and over the next.

Hip-Hip. Soldiers step over each bar; they either alternate legs or use the same lead leg each time.

Balancing Logs. Soldiers step up on a log and walk or run along it while keeping their balance.

Island Hopper. Soldiers jump from one log to another until the obstacle is negotiated.

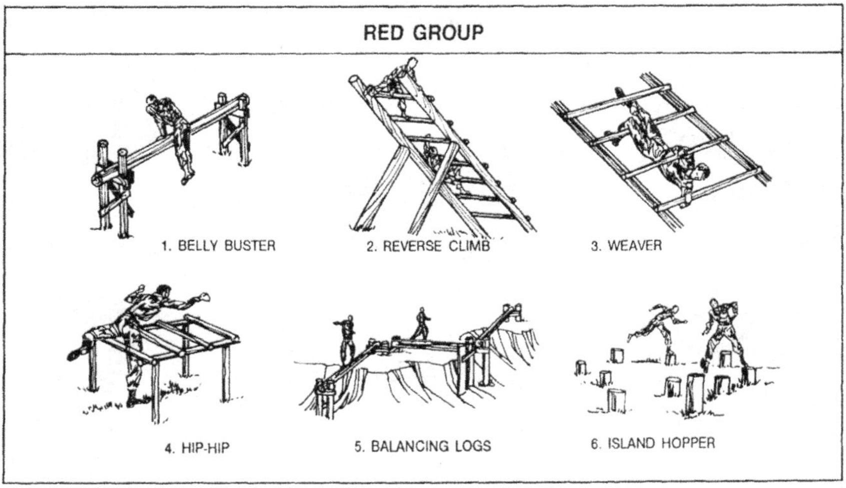

Figure 8-8

White Group

This group contains the second six obstacles. These are described below and numbered 7 through 12 in Figure 8-9.

Tough Nut. Soldiers step over each X in the lane.

Inverted Rope Descent. Soldiers climb the tower, grasp the rope firmly, and swing their legs upward. They hold the rope with their legs to distribute the weight between their legs and arms. Braking the slide with their feet and legs, they proceed down the rope. Soldiers must be warned that they may get rope burns on their hands. This obstacle can be dangerous when the rope is slippery. Soldiers leave the rope at a clearly marked point of release. Only one soldier at a time is allowed on the rope. Soldiers should not shake or bounce the ropes. This obstacle requires two instructors--one on the platform and the other at the base.

Low Belly-Over. Soldiers mount the low log and jump onto the high log. They grasp over the top of the log with both arms, keeping the belly area in contact with it. They swing their legs over the log and lower themselves to the ground.

Belly Crawl. Soldiers move forward under the wire on their bellies to the end of the obstacle. To reduce the tendency to push the crawling surface, it is filled with sand or sawdust to the far end of the obstacle. The direction of negotiating the crawl is reversed from time to time.

Easy Balancer. Soldiers walk up one inclined log and down the one on the other side to the ground.

Tarzan. Soldiers mount the lowest log, walk the length of it, then each higher log until they reach the horizontal ladder. They grasp two rungs of the ladder and swing themselves into the air. They negotiate the length of the ladder by releasing one hand at a time and swinging forward, grasping a more distant rung each time.

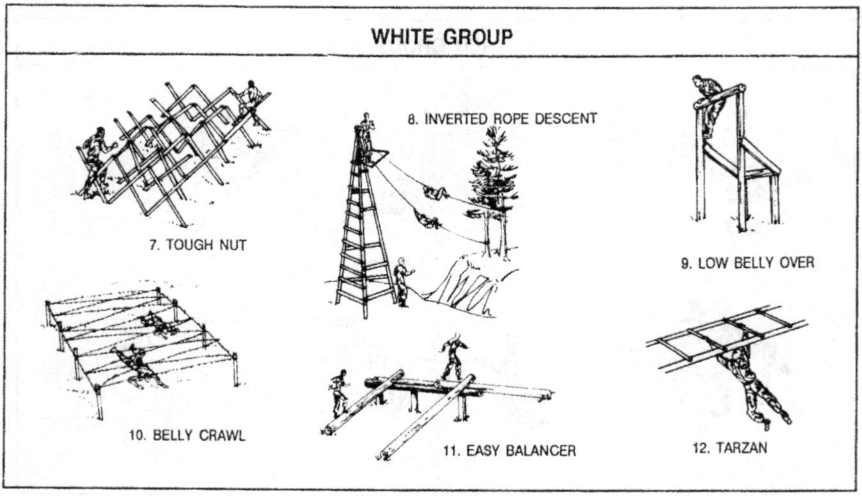

Figure 8-9

Blue Group

This group contains the third six obstacles. These are described below and numbered 13 through 18 in Figure 8-10.

High **Step-over. Soldiers step over** each log while alternating their lead foot or using the same one.

Swinger. Soldiers climb over the swing log to the ground on the opposite side.

Low Wire. Soldiers move under the wire on their backs while raising the wire with their hands to clear their bodies. To reduce the tendency to push the crawling surface, it is filled with sand or sawdust to the far end of the obstacle. The direction of negotiating the obstacle is alternated.

Swing, Stop, and Jump. Soldiers gain momentum with a short run, grasp the rope, and swing their bodies forward to the top of the wall. They release the rope while standing on the wall and jump to the ground.

Six Vaults. Soldiers vault over the logs using one or both hands.

Wall Hanger. Soldiers walk up the wall using the rope. From the top of the wall, they grasp the bar and go hand-over-hand to the rope on the opposite end. They use the rope to descend,

BLUE GROUP

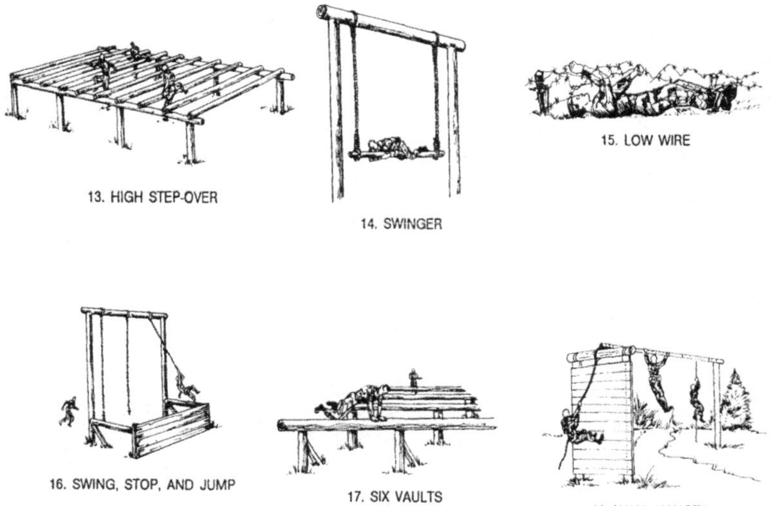

13. HIGH STEP-OVER

14. SWINGER

15. LOW WIRE

16. SWING, STOP, AND JUMP

17. SIX VAULTS

18. WALL HANGER

Figure 8-10

Black Group

This group contains the last six obstacles. These are described below and numbered 19 through 24 in Figure 8-11.

Inclining Wall. Soldiers approach the underside of the wall, jump up and grasp the top, and pull themselves up and over. They slide or jump down the incline to the ground.

Skyscraper. Soldiers jump or climb to the first floor and either climb the corner posts or help one another to the higher floors. They descend to the ground individually or help one another down. The top level or roof is off limits, and the obstacle should not be overloaded. A floor must not become so crowded that soldiers are bumped off. Soldiers should not jump to the ground from above the first level.

Jump and Land. Soldiers climb the ladder to the platform and jump to the ground.

Confidence Climb. Soldiers climb the inclined ladder to the vertical ladder. they go to the top of the vertical ladder, then down the other side to the ground.

Belly Robber. Soldiers step on the lower log and take a prone position on the horizontal logs. They crawl over the logs to the opposite end of the obstacle. Rope gaskets must be tied to the ends of each log to keep the hands from being pinched and the logs from falling.

The Tough One. Soldiers climb the rope or pole on the lowest end of the obstacle. They go over or between the logs at the top of the rope. They move across the log walkway, climb the ladder to the high end, then climb down the cargo net to the ground.

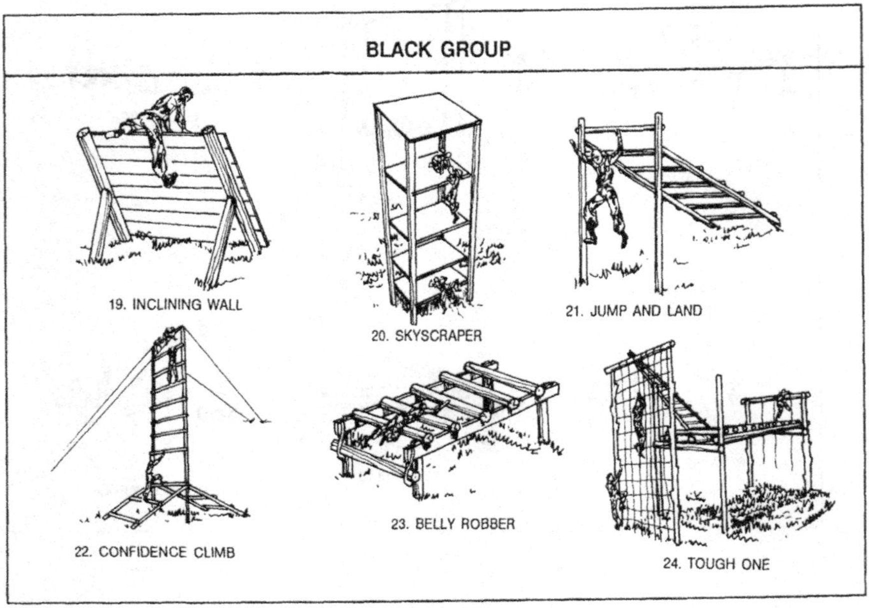

Figure 8-11

Rifle Drills

Rifle drills are suitable activities for fitness training while bivouacking or during extended time in the field. In most situations, the time consumed in drawing weapons makes this activity cumbersome for garrison use. However, it is a good conditioning activity, and the use of individual weapons in training fosters a warrior's spirit.

There are four rifle-drill exercises that develop the upper body. They are numbered in a set pattern. The main muscle groups strengthened by rifle drills are those of the arms, shoulders, and back.

Rifle drill is a fast-moving method of exercising that soldiers can do in as little as 15 minutes. With imagination, the number of steps and/or rifle exercises can be expanded beyond those described here.

EXERCISE PROGRESSION

The rifle-drill exercise normally begins with six repetitions and increases by one repetition for each three periods of exercise. This rate continues until soldiers can do 12 repetitions. However, the number of repetitions can be adjusted as the soldiers improve.

In exercises that start from the rifle-downward position, on the command "Move," soldiers execute port arms and assume the starting position. At the end of the exercise, the command to return soldiers to attention is "Position of attention, move."

In exercises that end in other than the rifle-downward position, soldiers assume that position before executing port arms and order arms.

These movements are done without command and need not be precise. Effective rifle exercises are strenuous enough to tire the arms. When the arms are tired, moving them with precision is difficult.

RIFLE DRILL EXERCISES

The following exercises are for use in rifle drills.

Up and Forward

This is a four-count exercise done at a fast cadence. (See Figure 8-12.)

Fore-Up, Squat

This is a four-count exercise done at a moderate cadence. (See Figure 8-13.)

Fore-Up, Behind Back

This is a four-count exercise done at a moderate cadence. (See Figure 8-14.)

Fore-Up, Back Bend

This is a four-count exercise done at moderate cadence. (See Figure 8- 15.)

UP AND FORWARD

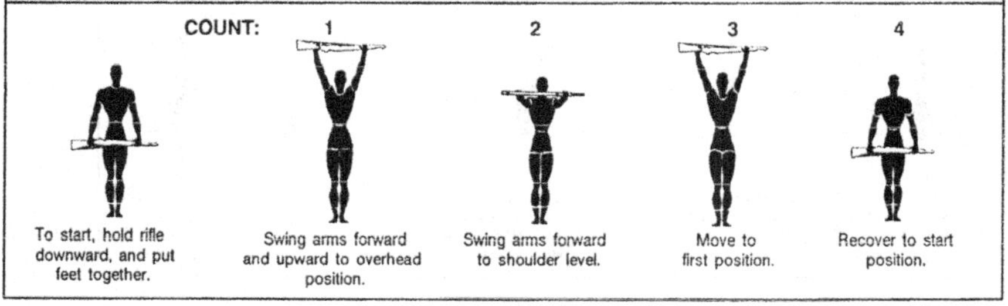

Figure 8-12

FORE-UP, SQUAT

COUNT:	1	2	3	4
To start, hold rifle downward, and put feet about shoulder width apart.	Swing arms forward and upward to overhead position.	Swing arms down to shoulder level, and assume half-knee-bend position.	Move to first position.	Recover to start position.

Figure 8-13

FORE UP, BEHIND BACK

COUNT:	1	2	3	4
To start, hold rifle downward, and put feet together.	Swing arms forward and upward to overhead position. Exhale.	Lower rifle to back of shoulders. Inhale.	Move to first position. Exhale.	Recover to start position. Inhale.

Figure 8-14

FORE-UP, BACK BEND

COUNT:	1	2	3	4
To start, hold rifle downward, and put feet together.	Swing arms forward and upward to overhead position.	Bend backward taking care not to bend too far. Keep face up and knees straight.	Move to first position.	Recover to start position.

Figure 8-15

Log Drills

Log drills are excellent for developing strength and muscular endurance, because they require the muscles to contract under heavy loads.

Log drills are team-conditioning exercises. They are excellent for developing strength and muscular endurance because they require the muscles to contract under heavy loads. They also develop teamwork and add variety to the PT program.

Log drills consist of six different exercises numbered in a set pattern. The drills are intense, and teams should complete them in 15 minutes. The teams have six to eight soldiers per team. A principal instructor is required to teach, demonstrate, and lead the drill. He must be familiar with leadership techniques for conditioning exercises and techniques peculiar to log drills.

AREA AND EQUIPMENT

Any level area is good for doing log drills. All exercises are done from a standing position. If the group is larger than a platoon, an instructor's stand may be needed.

The logs should be from six to eight inches thick, and they may vary from 14 to 18 feet long for six and eight soldiers, respectively. The logs should be stripped, smoothed, and dried. The 14-foot logs weigh about 300 pounds, the 18-foot logs about 400 pounds. Rings should be painted on the logs to show each soldier's position. When not in use, the logs are stored on a rack above the ground.

FORMATION

All soldiers assigned to a log team should be about the same height at the shoulders. The best way to divide a platoon is to have them form a single file or column with short soldiers in front and tall soldiers at the rear. They take their positions in the column according to shoulder height, not head height. When they are in position, they are divided into teams of six or eight. The command is "Count off by sixes (or eights), count off." Each team, in turn, goes to the log rack, shoulders a log, and carries it to the exercise area.

The teams form columns in front of the instructor. Holding the logs in chest position, they face the instructor and ground the log. Ten yards should separate log teams within the columns. If more than one column is used, 10 yards should separate columns.

STARTING DOSAGE AND PROGRESSION

The starting session is six repetitions of each exercise. The progression rate is an increase of one repetition for each three periods of exercise. Soldiers continue this rate until they do 12 repetitions with no rest between exercises. This level is maintained until another drill is used.

START POSITIONS

The soldiers fall in facing their log, with toes about four inches away. Figure 8-16 shows the basic starting positions and commands.

Right-Hand Start Position, Move

On the command "Move," move the left foot 12 inches to the left, and lower the body into a flatfooted squat. Keep the back straight, head up, and arms between the legs. Encircle the far side of the log with the left hand. Place the right hand under the log. (See 1, Figure 8-16.)

Left-Hand Start Position, Move

This command is done the same way as the preceding command. However, the left hand is under the log, and the right hand encircles its far side. (See 2, Figure 8-16.)

Right-Shoulder Position, Move

This command is given from the right-hand-start position. On the command "Move," pull the log upward in one continuous motion to the right shoulder. At the same time, move the left foot to the rear and stand up, facing left. Balance the log on the right shoulder with both hands. (See 3, Figure 8-16.) This movement cannot be done from the left-hand-start position because of the position of the hands.

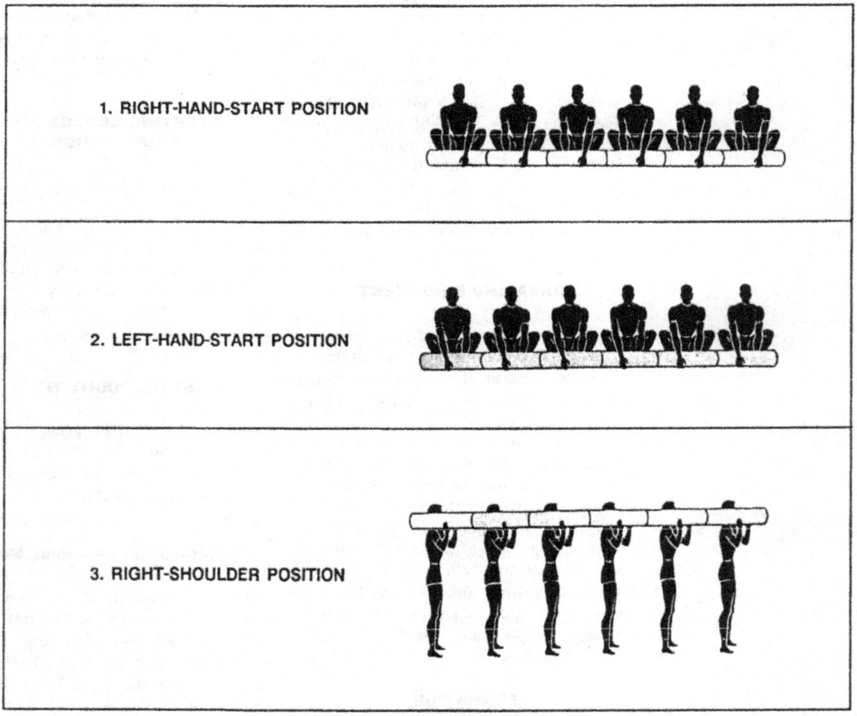

Figure 8-16

Left-Shoulder Position, Move

This command is given from the left-hand-start position. On the command "Move, " pull the log upward to the left shoulder in one continuous motion. At the same time, move the right foot to the rear, and stand up facing right. Balance the log on the left shoulder with both hands. (See 4, Figure 8-17.) This movement cannot be done from the right-hand-start position.

Waist Position, Move

From the right-hand-start position, pull the log waist high. Keep the arms straight and fingers laced under the log. The body is inclined slightly to the rear, and the chest is lifted and arched. (See 5, Figure 8-17.)

Chest Position, Move

This command is given after taking the waist position. On the command "Move," shift the log to a position high on the chest, bring the left arm under the log, and hold the log in the bend of the arms. (See 6, figure 8-17.) Keep the upper arms parallel to the ground.

To move the log from the right to the left shoulder, the command is "Left-shoulder position, move." Push the log overhead, and lower it to the opposite shoulder.

To return the log to the ground from any of the above positions, the command is "Start position, move." At the command "Move," slowly lower the log to the ground. Position the hands and fingers so they are not under the log.

4. LEFT-SHOULDER POSITION

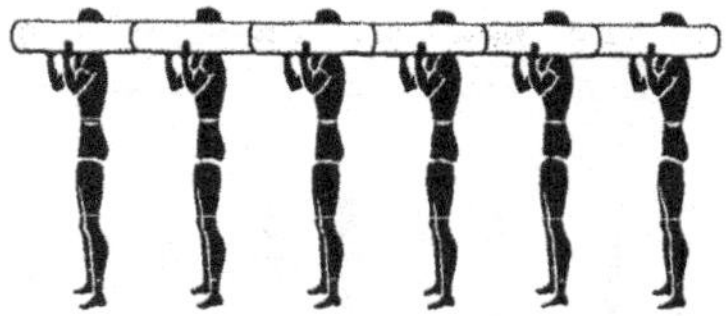

5. WAIST POSITION

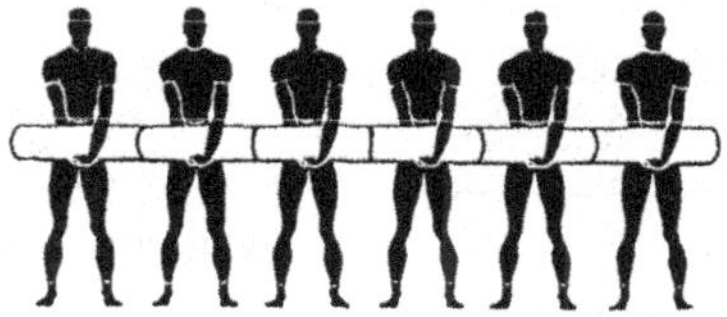

6. CHEST POSITION

Figure 8-17

LOG-DRILL EXERCISES

The following are log-drill exercises.

Exercise 1. Two-Arm Push-Up

Start Position: Right- or left-shoulder position, with feet about shoulder-width apart. (See 1, Figure 8-18.)

Cadence: Moderate.

Movement: A four-count exercise; at the count of --

"One"-Push the log overhead until the elbows lock.

"Two"-Lower the log to the opposite shoulder.

"Three"-Repeat the action of count one.

"Four"-Recover to the start position.

Exercise 2. Forward Bender

Start Position: Chest position, with feet about shoulder-width apart. (See 2, Figure 8-18.)

Cadence: Moderate.

Movement A four-count exercise; at the count of --

"One"-Bend forward at the waist while keeping the back straight and the knees slightly bent.

"Two"-Recover to the start position.

'Three"-Repeat the action of count one.

"Four"-Recover to the start position.

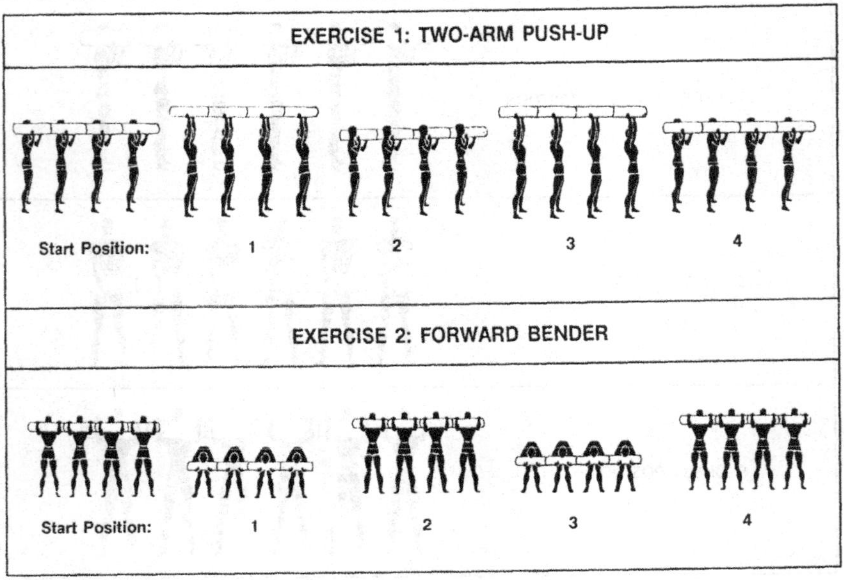

Figure 8-18

Exercise 3. Straddle Jump

Start Position Right- or left-shoulder position, with feet together, and fingers locked on top of the log. Pull the log down with both hands to keep it from bouncing on the shoulder. (See 3, Figure 8-19.)

Cadence: Moderate.

Movement A four-count exercise; at the count of--

"One''-Jump to a side straddle.

"Two''-Recover to the start position.

'Three''-Repeat the action of count one.

"Four''-Recover to the start position.

Exercise 4. Side Bender

Start Position: Right-shoulder position with the feet about shoulder-width apart. (See 4, Figure 8- 19.)

Cadence Moderate.

Movement: A four-count exercise; at the count of--

"One''-Bend sideward to the left as far as possible, bending the left knee.

"Two''-Recover to the start position.

"Three''-Repeat the action of count one.

"Four''-Recover to the start position.

NOTE: After doing the required number of repetitions, change shoulders and do an equal number to the right side.

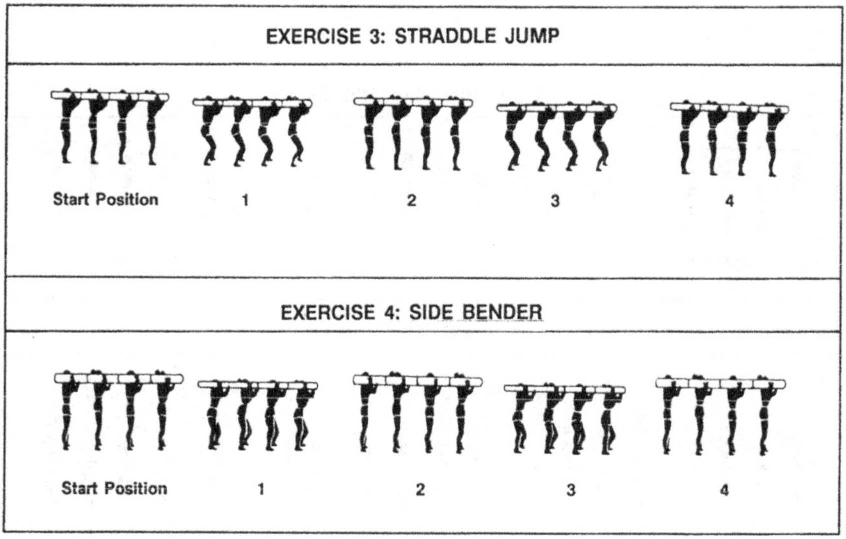

Figure 8-19

Exercise 5. Half-Knee Bend

Start Position: Right- or left-shoulder position, with feet about shoulder-width apart, and fingers locked on top of the log. (See 5, Figure 8-20.)

Cadence: Slow.

Movement: A four-count exercise; at the count of --

"One''-Flex the knees to a half-knee bend.

"Two''-Recover to the start position.

"Three''-Repeat the action of count one.

"Four''-Recover to the start position.

(NOTE: Pull forward and downward on the log throughout the exercise.)

Exercise 6. Overhead Toss (NOTE: Introduce this exercise only after soldiers have gained experience and strength by doing the other exercises for several sessions.)

Start Position: Right-shoulder position with the feet about shoulder-width part. The knees are at a quarter bend. (See 6, Figure 8-20.)

Cadence: Moderate.

Movement: A four-count exercise; at the count of --

"One''-Straighten the knees and toss the log about 12 inches overhead. Catch the log with both hands, and lower it toward the opposite shoulder. As the log is caught, lower the body into a quarter bend.

"Two''-Again, toss the log into the air and, when caught, return it to the original shoulder.

"Three''-Repeat the action of count one.

"Four''-Recover to the start position.

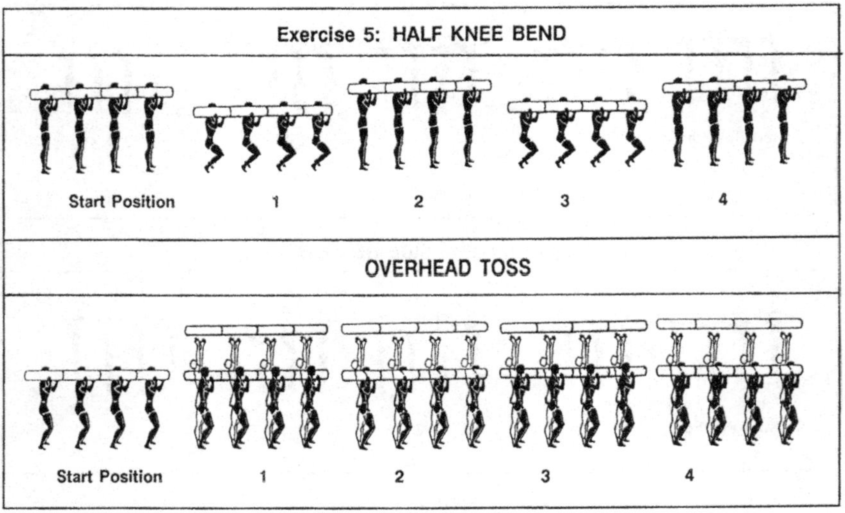

Figure 8-20

Aquatic Exercise

Aquatics is a mode of physical training which helps one attain and maintain physical fitness through exercises in the water. It is sometimes called slimnastics. Aquatic training can improve muscular endurance, CR endurance, flexibility, coordination, and muscular strength.

Because of its very low impact to the body, an aquatic exercise program is ideal for soldiers who are overweight and those who are limited due to painful joints, weak muscles, or profiles. The body's buoyancy helps minimize injuries to the joints of the lower legs and feet. It exercises the whole body without jarring the bones and muscles. Leaders can tailor the variety and intensity of the exercises to the needs of all the soldiers in the unit.

Aquatic training is a good supplement to a unit's PT program. Not only is it fun, it exposes soldiers to water and can make them more comfortable around it. Most Army installations have swimming pools for conducting aquatic, physical training sessions.

SAFETY CONSIDERATIONS

One qualified lifeguard is needed for every 40 soldiers at all aquatic training sessions. Nonswimmers must remain in the shallow end of the pool. They should never exercise in the deep end with or without flotation devices.

EQUIPMENT

Soldiers normally wear swim suits for aquatics, but they can wear boots and fatigues to increase the intensity of the activities. The following equipment is optional for training:

- Goggles.
- Kickboard.
- Pull buoy.
- Ear/nose plugs.
- Fins.
- Hand paddles.

SAMPLE TRAINING PROGRAM

'Warm-Up

As in any PT session, a warm-up is required. It can be done in the water or on the deck. Allow five to seven minutes for the warm-up.

Conditioning Phase

Soldiers should exercise vigorously to get a training effect. Energetic music may be used to keep up the tempo of the workout. The following are some exercises that can be used in an aquatic workout. (See Figure 8-21.)

Side Leg-Raises. Stand in chest to shoulder-deep water with either side of the body at arm's length to the wall of the pool, and grasp the edge with the nearest hand. Raise the outside leg sideward and upward from the hip. Next, pull the leg down to the starting position. Repeat these actions. Then, turn the other side of the body to the wall, and perform the exercise with the other leg. DURATION: 30 seconds (15 seconds per leg).

Leg-Over. Stand in chest-to shoulder-deep water, back facing the wall of the pool. Reach backward with the arms extended, and grasp the pool's edge. Next, raise one leg in front of the body away from the wall, and move it sideward toward the other leg as far as it can go. Then, return the leg to the front-extended position, and lower it to the starting position. Repeat these actions with the other leg, and continue to alternate legs. DURATION: 30 seconds (15 seconds per leg).

Rear Leg Lift. Stand in chest-to shoulder-deep water with hands on the pool's edge, chest to the wall. Raise one leg back and up from the hip, extend it, and point the foot. Then, pull the leg back to the starting position. Alternate these actions back and forth with each leg. DURATION: 20 seconds (10 seconds each leg).

AN AQUATIC EXERCISE WORKOUT CENTER

Figure 8-21

Alternate Toe Touch. Stand in waist-deep water. Raise the left leg as in kicking while touching the elevated toe with the right hand. At the same time, rotate the head toward the left shoulder, and push the left arm backward through the water. Alternate these actions back and forth with each leg and opposite hand. DURATION 2 minutes.

Side Straddle Hop. Stand in waist-deep water with hands on hips and feet together. Jump sideward and land with feet about two feet apart. Then, return to the starting position, and repeat the jumping action. DURATION 2 minutes.

Stride Hop. Stand in waist-deep water with hands on hips and feet together. Jump, moving the left leg forward and right leg backward. Then, jump again moving the right leg forward and left leg backward. Repeat these actions. DURATION 2 minutes.

The Bounce. Stand in waist-deep water with hands on hips and feet together. Jump high with feet together. Upon landing, use a bouncing motion, and repeat the action. DURATION: 1 minute.

Rise on Toes. Stand in chest-to shoulder-deep water with arms at sides and feet together. Rise up using the toes. Then, lower the body to the starting position. Repeat the action. DURATION: 1 minute.

Side Bender. Stand in waist-deep water with the left arm at the side and the right arm extended straight overhead. Stretch slowly, bending to the left. Recover to the starting position, and repeat the action. Next, reverse to the right arm at the side and the left arm extended straight overhead. Repeat the stretching action to the right side. DURATION: 1 minute.

Walking Crawl. Walk in waist- to chest-deep water. Simulate the overhand crawl stroke by reaching out with the left hand cupped and pressing the water downward to the thigh. Repeat the action with the right hand. Alternate left and right arm action. DURATION: 2 minutes.

Bouncing. Stand in chest-deep water, arms at sides. Bounce on the left foot while pushing down vigorously with both hands. Repeat the action with the right foot. Alternate bouncing on the left and right foot. DURATION: 2 minutes.

Bounding in Place with Alternate Arm Stretch, Forward. Bound in place in waist-deep water using high knee action. Stretch the right arm far forward when the left knee is high and the left arm is stretched backward. When the position of the arm is reversed, simulate the action of the crawl stroke by pulling down and through the water with the hand. DURATION 1 minute.

Poolside Knees Up, Supine. Stand in chest-to shoulder-deep water, back against the wall of the pool. Extend the arms backward, and grasp the pool's edge. With feet together, extend the legs in front of the torso, and assume a supine position. Then with the legs together, raise the knees to the chin. Return to the starting position, and repeat the action. DURATION: 2 minutes (maximum effort).

Twisting Legs, Supine. Stand in chest-to shoulder-deep water, back against the wall of the pool. Extend the arms backward, and grasp the pool's edge. With feet together, extend the legs in front of the torso, and assume a supine position. Then, twist the legs slowly to the left, return to the starting position, and twist the legs slowly to the right. Repeat this twisting action. DURATION: 1 minute (2 sets, 30 seconds each).

Scissor Kick. Float in chest- to shoulder- deep water on either side of the body with the top arm extended, hand holding the pool's edge. Brace the bottom hand against the pool's wall with feet below the water's surface. Next, assume a crouching position by gringing the heels toward the hips by bending the knees. Then, straighten and spread the legs with the top leg extending backward. When the legs are extended and spread, squeeze them back together (scissoring). Pull with the top hand, and push with the bottom hand. The propulsive force of the kick will tend to cause the body to rise to the water's surface. DURATION 1 minute (2 sets, 30 seconds each, maximum effort).

Push Away. Stand in chest-to shoulder-deep water facing the pool's wall and at arm's length from it. Grasp the pool's edge, and bend the arms so that the body is leaning toward the wall of the pool. Vigorously push the chest back from the wall by straightening the arms. Then, with equal vigor, pull the upper body back to the wall. Repeat these actions. DURATION: 2 minutes (maximum effort).

Gutter Push-Ups. Stand in chest-to shoulder- deep water facing the pool's wall. Place the hands on the edge or gutter of the pool. Then, raise the body up and out of the water while extending the arms. repeat this action. DURATION: 2 minutes (4 sets, 30 seconds each with 5-second rests between sets).

Front Flutter Kick. Stand in chest-to shoulder-deep water facing the pool's wall. Grasp the pool's edge or gutter and assume a prone position with legs extended just below the water's surface. Then, kick flutter style, toes pointed, ankles flexible, knee joint loose but straight. The legs should simulate a whip's action. DURATION 1 minute (2 sets, 30 seconds each).

Running. Move in a running gait in chest-to shoulder-deep water with arms and hands under the water's surface. This activity can be stationary, or the exerciser may run from poolside to poolside. Runners must concentrate on high knee action and good arm movement. DURATION 10 to 20 minutes.

The Engine. Stand in chest-to shoulder-deep water, arms straight and in front of the body and parallel to the water with the palms facing downward. While walking forward, raise the left knee to the left elbow, then return to the starting position. Continuing to walk forward, touch the right knee to the right elbow, and return to the starting position. Be sure to keep the arms parallel to the water throughout the exercise. DURATION 1 to 2 minutes (2 sets).

Cool-Down

This is required to gradually bring the body back to its pre-exercise state. It should last from five to seven minutes.

CHAPTER 9

Competitive Fitness Activities

Physical fitness is one of the foundations of combat readiness, and maintaining it must be an integral part of every soldier's life. This chapter discusses competitive fitness activities and athletic events that commanders can use to add variety to a unit's physical fitness program. There is also a section on developing a unit intramural program. Athletic and competitive fitness activities are sports events which should only be used to supplement the unit's PT program. They should never replace physical training and conditioning sessions but, rather, should exist to give soldiers a chance for healthy competition. Only through consistent, systematic physical conditioning can the fitness components be developed and maintained.

Crucial to the success of any program is the presence and enthusiasm of the leaders who direct and participate in it. The creativity of the physical training planners also plays a large role. Competitive fitness and athletic activities must be challenging. They must be presented in the spirit of fair play and good competition.

It is generally accepted that competitive sports have a tremendous positive influence on the physical and emotional development of the participants. Sports competition can enhance a soldier's combat readiness by promoting the development of coordination, agility, balance, and speed. Competitive fitness activities also help develop assets that are vital to combat effectiveness. These include team spirit, the will to win, confidence, toughness, aggressiveness, and teamwork.

Competitive fitness activities help in the development of assets that are vital to combat effectiveness.

Intramural

The Army's sports mission is to give all soldiers a chance to participate in sports activities. A unit-level intramural program can help achieve this important goal. DA Pam 28-6 describes how to organize various unit-level intramural programs.

Factors that affect the content of the sports program differ at every Army installation and unit. Initiative and ingenuity in planning are the most vital assets. They are encouraged in the conduct of every program.

OBJECTIVES

A well-organized and executed intramural program yields the following:

- Team spirit, the will to win, confidence, aggressiveness, and teamwork. All are vital to combat effectiveness.
- A change from the routine PT program.
- The chance for all soldiers to take part in organized athletics.

ORGANIZATION

The command level best suited to organize and administer a broad intramural program varies according to a unit's situation. If the objective of maximum participation is to be achieved, organization should start at company level and then provide competition up through higher unit levels. Each command level should have its own program and support the next higher program level.

To successfully organize and conduct an intramural program, developers should consider the following factors and elements.

Authority

The unit commander should publish and endorse a directive giving authorization and guidance for a sports program. A detailed SOP should also be published.

Personnel

Leaders at all levels of the intramural program should plan, organize, and supervise it. Appointments at all echelons should be made for at least one year to provide continuity. The commander must appoint a qualified person to be the director, regardless of the local situation, type, and size of the unit. The director must be a good organizer and administrator and must have time to do the job correctly. He should also have a sense of impartiality and some athletic experience.

Commanders should form an intramural sports council in units of battalion size or larger and should appoint members or require designated unit representatives. The council should meet at least once a month or as often as the situation requires. The council serves as an advisory body to the unit commander and intramural director. It gives guidance about the organization and conduct of the program.

Facilities and Equipment

Adequate facilities and equipment must be available. When facilities are limited, leaders must plan activities to ensure their maximum use. In all cases, activities must be planned to ensure the safety of participants and spectators.

Funds and Budget

Adequate funds are essential to successfully organize and operate a sports program. Therefore, beforehand, organizers must determine how much money is available to support it. To justify requests for funds they must prepare a budget in which they justify each sports activity separately. The budget must include special equipment, supplies, awards, pay for officials, and other items and services. Units can reduce many of their costs by being resourceful.

AWARD SYSTEM

Commanders can stimulate soldiers to participate in competitive athletics by using an award system.

Commanders can stimulate units and soldiers to participate in competitive athletics by using an award system. One type is a point-award system where teams get points based on their win/loss records and/or final league standings. This reflects the unit's standings in the overall intramural sports program. The recognition will help make units and individuals participate throughout the year. Trophies can then be given for overall performance and individual activities.

PROGRAM PLANNING

A successful program depends on sound plans and close coordination between the units involved. The intramural director should meet with subordinate commanders or a sports representative to determine what program of activities is compatible with the mission and training activities of each unit. Unless they resolve this issue, they may not get command support which, in turn, could result in forfeitures or lack of participation. The less-popular activities may not be supported because of a lack of interest.

Evaluations

Before the program is developed, leaders must study the training and availability situation at each unit level. They should include the following items in a survey to help them determine the scope of the program and to develop plans:

- ***General.*** Evaluate the commander's attitude, philosophy, and policy about the sports program. Understand the types of units to be served, their location, the climate, and military responsibilities.
- ***Troops.*** Determine the following: 1) number and types of personnel; 2) training status and general duty assignment; 3) special needs, interests, and attitudes.
- ***Time available.*** Coordinate the time available for the sports program with the military mission. Determine both the on-duty and off-duty time soldiers have for taking part in sports activities.
- ***Equipment.*** Consider the equipment that will be needed for each sport.
- ***Facilities.*** Determine the number, type, and location of recreational facilities both within the unit and in those controlled by units at higher levels.
- ***Funds.*** Determine how much each unit can spend on the intramural program.
- ***Personnel.*** Assess how many people are needed to run the program. The list should include a director and assistants, sports council, officials, and team captains, as well as volunteers for such tasks as setting up a playing field.
- ***Coordination.*** Coordinate with the units' operations sections to avoid conflict with military training schedules.
- ***Activities.*** The intramural director should plan a tentative program of activities based on the season, local situation, and needs and interests of the units. Both team and individual sports should be included. Some team sports are popular at all levels and need little promotional effort for success. Among these are volleyball, touch football, basketball, and softball. Some individual competitive sports have direct military value. They include boxing, wrestling, track and field, cross country, triathlon, biathlon, and swimming. While very popular, these sports are harder to organize than team sports. See Figures 9-1 and 9-2 for a list of sports activities.

SPORTS ACTIVITIES			
Team Sports			
Baseball Flag Football Softball	Basketball Water Polo Speedball	Field Hockey Pushball Tug-of-War	Football Soccer Volleyball

Field-Type-Meets		
Athletic Carnivals Physical Fitness Meet Track and Field Urban Orienteering	Cross Country Relay Carnival Water Carnival	Military Field Meets Swimming and Diving Unit Olympics

Figure 9-1

SPORTS ACTIVITIES			
Individual Sports			
Archery Boxing Handball Marathon Track & Field Triathlon	Badminton Canoeing Judo Squash Rowing Skating	Tennis Table Tennis Horseshoes Skating Sky Diving Weightlifting	Bowling Gymnastics Modern Biathlon Mountain Climbing Skeet Shooting Swimming and Diving

Figure 9-2

ESSENTIAL ELEMENTS

Intramural Handbook

- Commander's foreward.
- Personnel directory.
- Title page.
- Purpose.
- References.
- Objectives.
- Duties of the personnel.
- Eligibility rules.
- Intramural sports council.
- Protest and sportsmanship board.
- Budgets and funding.
- Officials association.
- Master calendar of activities.
- Organization of leagues and units of competition.
- Command- points award system.
- Facilities and their hours of operation.
- Equipment regulations.
- Rules and regulations of each sport.
- Reporting time for competition.
- Postponement of contests.
- Protest procedures.
- Awards.
- Records and results.
- Bulletin boards and publicity.

Table 9-1

Functions

Once the evaluations have been made, the following functions should be performed:

- ***Make a handbook.*** An intramural handbook should be published at each level of command from installation to company to serve as a standing operating procedure (SOP). This handbook should include the essential elements listed in Table 9-1 above.
- ***Plan the calendar.*** Local situations and normal obstacles may conflict with the intramural program. How ever, a way can be found to provide a scheduled program for every season of the year.
- ***Choose the type of competition.*** Intramural directors should be able to choose the type of competition best suited for the sport and local circumstances. They should also know how to draw up tournaments. Unless the competition must take place in a short time, elimination tournaments should not be used. The round-robin tournament has the greatest advantage because individuals and teams are never eliminated. This type of competition is adaptable to both team and individual play. It is appropriate for small numbers of entries and league play in any sport.
- ***Make a printed schedule.*** Using scheduling forms makes this job easier. The form should include game number, time, date, court or field, and home or visiting team. Space for scores and officials is also helpful. Championship games or matches should be scheduled to take place at the best facility.

Unit Activities

The following games and activities may be included in the unit's PT program. They are large-scale activities which can combine many components of physical and motor fitness. In addition, they require quick thinking and the use of strategy. When played vigorously, they are excellent activities for adding variety to the program.

NINE-BALL SOCCER

The object of this game is for each of a team's five goalies to have one ball.

Players

There are 25 to 50 players on each team, five of whom are goalies. The other players are divided into four equal groups. The goalies play between the goal line and 5-yard line of a standard football field. The other four groups start the game between the designated 10-yard segments of the field. (See Figure 9-3.) The goalies and all other players must stay in their assigned areas throughout the game. The only exceptions are midfielder who stand between the 35- and 45-yard lines. These players may occupy both their assigned areas and the 10-yard free space at the center of the field.

The Game

The game starts with all players inside their own areas and midfielder on their own 40-yard line. The nine balls are placed as follows. Four are on each 45-yard line with at least five yards between balls. One is centered on the 50-yard line. The signal to start play is one long whistle blast. Players must pass the balls through the opposing team's defenses into the goal area using only their feet or heads. The first team whose goalies have five balls wins a point. The game then stops, and the balls are placed for the start of a new set. The first team to score five points wins.

There are no time-outs except in case of injury, which is signaled by two sharp whistle blasts. The teams change positions on the field after each set. Team members move to different zones after the set.

Rules

A ball is played along the ground or over any group or groups of players. The ball may travel any distance if it is played legally.

Goalies may use their hands in playing the ball and may give a ball to other goalies on their team. For a set to officially end, each goalie must have a ball.

If players engage in unnecessary roughness or dangerous play, the referee removes them from the game for the rest of the set and one additional set. He also removes players for the rest of the set if they step on or over a boundary or sideline or use their hands outside the goal area.

If a goalie steps on or over a boundary or sideline, the referee takes the ball being played plus another ball from the goalie's team and gives these balls to the nearest opposing player. If

NINE-BALL SOCCER

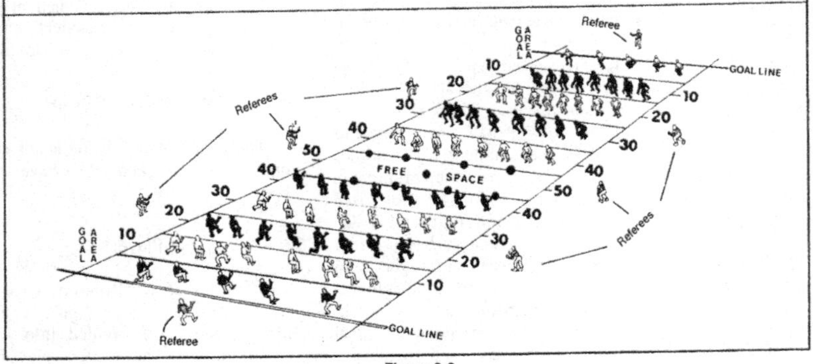

Figure 9-3

the team has no other ball in the goal area, the referee limits the penalty to the ball that is being played.

If a ball goes out of bounds, the referee retrieves it. The team that caused it to go out of bounds or over the goal line loses possession. The referee puts the ball back into play by rolling it to the nearest opposing player.

PUSHBALL

This game requires a large pushball that is five to six feet in diameter. It also requires a level playing surface that is 240 to 300 feet long and 120 to 150 feet wide. The length of the field is divided equally by a center line. Two more lines are marked 15 feet from and parallel to the end lines and extending across the entire field. (See Figure 9-4.)

Players

There are 10 to 50 soldiers on each of two teams.

The Game

The object of the game is to send the ball over the opponent's goal line by pushing, rolling, passing, carrying, or using any method other than kicking the ball.

The game begins when the ball is placed on the centerline with the opposing captains three feet away from it. The other players line up 45 feet from the ball on their half of the field. At the referee's starting whistle, the captains immediately play the ball, and their teams come to their aid.

At quarter time, the ball stays dead for two minutes where it was when the quarter ended. At halftime, the teams exchange goals, and play resumes as if the game were beginning.

A team scores a goal when it sends the ball across the opposing team's end line. A goal counts five points. The team that scores a goal may then try for an extra point. For the extra point, the ball is placed on the opposing team's 5-yard line, and the teams line

PUSHBALL

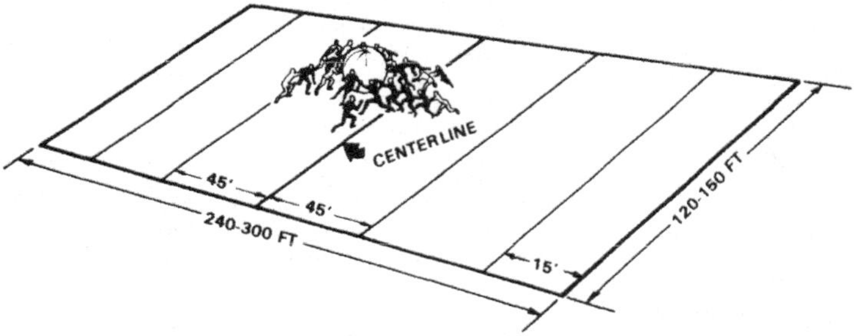

Figure 9-4

up across the field separated by the width of the ball. Only one player may place his hands on the ball. The player who just scored is directly in front of the ball. At the referee's signal, the ball is put into play for one minute. If any part of the ball is driven across the goal line in this period, the offense scores one point. The defense may not score during the extra point attempt.

The game continues until four 10-minute quarters have been played. Rest periods are allowed for two minutes between quarters and five minutes at halftime.

Rules

Players may use any means of interfering with the opponents' progress except striking and clipping. Clipping is throwing one's body across the back of an opponent's legs as he is running or standing. Force may legally be applied to all opponents whether they are playing the ball or not. A player who strikes or clips an opponent is removed from the game, and his team is penalized half the distance to its goal.

When any part of the ball goes out of bounds, it is dead. The teams line up at right angles to the sidelines. They should be six feet apart at the point where the ball went out. The referee tosses the ball between the teams.

When, for any reason, the ball is tied up in one spot for more than 10 seconds, the referee declares it dead. He returns the ball into play the same way he does after it goes out of bounds.

STRATEGY PUSHBALL

Strategy pushball is similar to pushball except that it is played on two adjacent fields, and opposing teams supply soldiers to the games on both fields. Team commanders assess the situation on the fields and distribute their soldiers accordingly. The commander decides the number of soldiers used, within limits imposed by the rules. This number may be adjusted throughout the game. Play on both fields occurs at the same time, but each game progresses independently. At the end of play, a team's points from both fields are added together to determine the overall winner.

This game requires two pushballs that are five to six feet in diameter. Pull-over vests or jerseys of two different colors are used by each team for a total of four different colors. Starters and reserves should be easily distinguishable. Starters and substitutes should wear vests of one color, while the team commander and reserves wear vests of the second color.

Players may wear any type of athletic shoes except those with metal cleats. Combat boots may be worn, but extra caution must be used to prevent injuries caused by kicking or stepping on other players. Soldiers wearing illegal equipment may not play until the problem has been corrected.

The playing area is two lined-off fields. These are 240 to 300 feet long by 120 to 150 feet wide. They are separated lengthwise by a 20-foot-wide divider strip. The length of each field is divided equally by a centerline that is parallel to the goal lines. Lines are also marked 45 feet from each side of the centerline and parallel to it. The lines extend across both fields. Dimensions may be determined locally based on available space and the number of players. The space between the fields is the team area. Each team occupies the third of the team space that immediately adjoins its initial playing field.

Time periods should be adjusted to suit weather conditions and soldiers' fitness levels.

Players

There are 25 to 40 soldiers on each team. A typical, 25-member team has the following:

- One team commander. He is responsible for overall game strategy and for determining the number and positions of players on the field.
- Sixteen starting members. Eight are on each field at all times; one is appointed field captain.
- Four reserve members. These are players the team commander designates as reinforcements.
- Three substitutes. These are replacements for starters or reserves.
- One runner. He is designated to convey messages from the team commander to field captains.

The proportion of soldiers in each category stays constant regardless of the total number on a team. Before the event, game organizers must coordinate with participating units and agree on the number on each team.

Runners serve at least one period; they may not play during that period. They are allowed on the field only during breaks in play after a dead ball or goal.

Reserves are used at any point in the game on either field and are committed as individuals or groups. They may enter or leave the playing field at any time whether the ball is in play or not. Team commanders may enter the game as reserves if they see the need for such action.

Reserves, substitutes, and starting members may be redesignated into any of the other components on a one-for-one basis only during dead balls, injury time-outs, or quarter- and half-time breaks. A reserve may become a starter by switching vests with an original starter, who then becomes a reserve.

When possible, senior NCOS and officers from higher headquarters or other units should be used as officials. Players must not question an official's

STRATEGY PUSHBALL

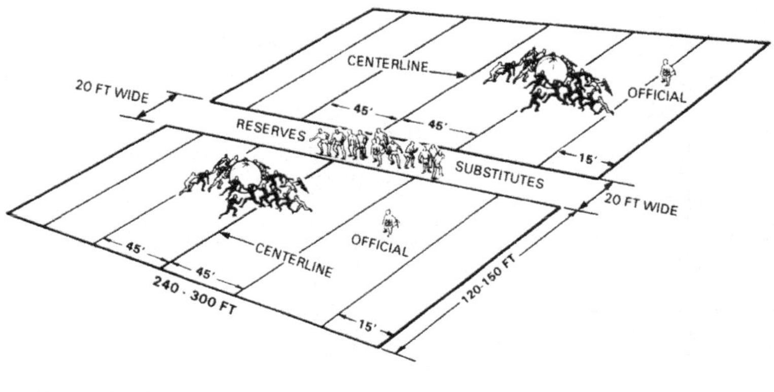

Figure 9-5

authority during play. Otherwise, the game can quickly get out of control.

Chain-of-command personnel should act as team commanders and field captains whenever possible.

The Game

The object is to propel the ball over the opponent's goal line by pushing, rolling, passing, carrying, or using any means other than kicking.

The game is officiated by two referees on each field, a chief umpire, and a scorekeeper. Referees concentrate on player actions so that they can quickly detect fouls and assess penalties. The chief umpire and scorekeeper occupy any area where they can best officiate the games. The chief umpire monitors the use of substitutes and reserves and ensures smooth progress of the games on both fields. The number of officials may be increased if teams have more than 25 players. Referees use their whistles to stop and start play except at the start and end of each quarter. The scorekeeper, who times the game with a stopwatch, starts and ends each quarter and stops play for injuries with some noisemaker other than a whistle. He may use such devices as a starter's pistol, klaxon, or air horn.

The game begins after the ball is placed on each field's center mark. Opposing field captains are three feet from the ball (six feet from the centerline). The rest of the starters are lined up 45 feet from the ball on their half of the field. (See Figure 9-5.) At the scorekeeper's signal, field captains immediately play the ball, and their teams come to their aid.

Starters may be exchanged between the fields if the minimum number of starters or substitutes per field is maintained.

Substitutes may enter the game only during breaks in play after a dead ball, goal, or time-out for injury. A substitute may not start to play until the player being replaced leaves the field.

When any part of the ball goes out of bounds, it is dead. The teams line up at right angles to the sidelines; they are 10 feet apart at the point where the ball went out of bounds. The referee places the ball between the teams at a point 15 feet inside the sideline. Play resumes when the referee blows the whistle.

When the ball gets tied up in one spot for more than 10 seconds for any reason, the referee declares it dead. He restarts play as with an out-of-bounds dead ball, except that he puts the ball on the spot where it was stopped.

Time does not stop for dead balls or goals. Play continues on one field while dead balls are restarted on the other.

At each quarter break, the ball stays on the spot where it was when the quarter ended. The next quarter, signaled by the scorekeeper, starts as it does after a ball goes out of bounds. At halftime the teams exchange goals, and play resumes as if the game were beginning.

A goal is scored when any part of the ball breaks the plane of the goal line between the sidelines. A goal counts one point. At the end of the fourth quarter, the points of each team from both fields are added together to determine the winner.

If there is a tie, a three-minute overtime is played. It is played the same as in regulation play, but only one field is used, with starting squads from both teams opposing each other. For control purposes, no more than 15 players per team are allowed on the field at once. The team with more points at the end of the overtime wins the game. If the game is still tied when time expires, the winner is the team that has gained more territory.

The game continues until four 10-minute quarters have been played. There is a 10-minute halftime between the second and third quarters. The clock stops at quarter breaks and halftime. Time-out is allowed only for serious injury. Play is then stopped on both fields.

Rules

Players may use any means of interfering with their opponents' progress, but they are penalized for striking or clipping opponents or throwing them to the ground. These penalties are enforced by the referees. Force maybe legally applied to any opponent whether or not they are playing the ball. Blocking is allowed if blockers stay on their feet and limit contact to the space between waist and shoulders. Blockers may not swing, throw, or flip their elbows or forearms. Tackling opposing soldiers who are playing the ball is allowed. The chief umpire or any referee may call infractions and impose penalties for unsportsmanlike conduct or personal fouls on either field. Penalties may also be called for infractions committed on the field or sidelines during playing time, quarter- and halftime breaks, and time-outs. Personal fouls are called for the following:

- Illegal blocking (below an opponent's waist).
- Clipping (throwing the body across the back of the opponent's legs as he is running or standing).
- Throwing an opponent to the ground (that is, lifting and dropping or slamming a player to the ground in stead of tackling cleanly).
- Spearing, tackling, or piling on an opponent who is already on the ground.
- Striking or punching with closed fist(s).
- Grasping an opponent's neck or head.
- Kicking.
- Butting heads.

Unsportsmanlike conduct is called for abusive or insulting language that the referee judges to be excessive and blatant. It is also called against a player on the sidelines who interferes with the ball or with his opponents on the field. A player who violates these rules should be removed from the game and made to run one lap around both playing fields. A penalized player leaves the team shorthanded until he completes the penalty lap and the next break in play occurs on the field from which he was removed. The penalized player or a substitute then enters the game. Referees and the chief umpire may, at their discretion, eject any player who is a chronic violator or who is judged to be dangerous to other players, Once ejected, the player must leave both the field of play and team area. Substitutes for ejected players may enter during the next break in play that follows a goal scored by either team. They enter on the field from which the players were ejected.

BROOM-BALL HOCKEY

This game is played on ice or a frozen field using hockey rules. Players wear boots with normal soles and carry broom-shaped sticks with which they hit the ball into the goals.

The object of this game is for teams to score goals through the opponent's defenses. Using only brooms, players pass the ball through the opposing team to reach its goal. The first team to score five points wins. Broom ball provides a good cardiorespiratory workout.

Players

There are 15 to 20 players on each team. One is a goalie and the others are divided into three equal groups. The goalie plays in the goal area of a standard soccer or hockey field or along the goal line if the two opposing goals are the same size. One soccer ball, or some other type of inflated ball, is used. The players need no padding.

The three groups begin the game in center field. All players must stay in their designated space throughout the game. A diagram of the field is shown at Figure 9-6.

The Game

The face-off marks the start of the game, the second half, and the restart of play after goals. Each half lasts 15 minutes. For the face-off, each player is on his own half of the field. All players, except the two centers, are outside the center circle. The referee places the ball in the center of the circle between the two centers. The signal to begin play is one long blast on the whistle. The ball must travel forward and cross the center circle before being played by another player. There are no time-outs except for injury. The time-out signal is two sharp whistle blasts.

Rules

All players, including goalies, must stay inside their legal boundaries at all times. Only goalies may use their hands to play the ball, but they must always keep control of their sticks. Other players must stay in their respective zones of play (Attack, Defense, Centerfield). The ball is played along the ground or over one or more groups of players. It may travel any distance as long as it is legally played.

The referee calls infractions and imposes penalties. Basic penalties are those called for the following:

- Unnecessary roughness or dangerous play. (The player is removed from the game; he stays in the penalty box for two minutes.)
- Ball out-of-bounds. (The team that caused it to go out loses possession, and the opposing team puts the ball back into play by hitting it to the nearest player.)
- Use of hands by a player other than a goalie. (The player must stay in the penalty box one minute.)
- Improper crossing of boundaries. (When a member of the team in possession of the ball crosses the bound ary line of his zone of play, possession will be awarded to the other team.)

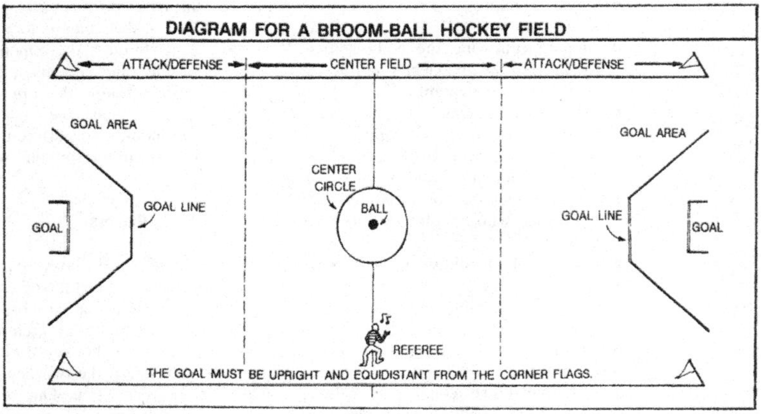

Figure 9-6

Orienteering combines map reading, compass use, and terrain study with strategy, competition, and exercise.

Orienteering

Orienteering is a competitive form of land navigation. It combines map reading, compass use, and terrain study with strategy, competition, and exercise. This makes it an excellent activity for any training schedule.

An orienteering course is set up by placing control points or marker signs over a variety of terrain. The orienteer or navigator uses a detailed topographical map and a compass to negotiate the course. The map should be 1:25,000 scale or larger. A liquid-filled orienteering compass works best. The base of the compass is transparent plastic, and it gives accurate readings on the run. The standard military, lensatic compass will work even though it is not specifically designed for the sport.

The best terrain for an orienteering course is woodland that offers varied terrain. Several different courses can be setup in an area 2,000 to 4,000 yards square. Courses can be short and simple for training beginners or longer and more difficult to challenge the advanced competitors.

The various types of orienteering are described below.

CROSS-COUNTRY ORIENTEERING

This popular type of orienteering is used in all international and championship events. Participants navigate to a set number of check or control points in a designated order. Speed is important since the winner is the one who reaches all the control points in the right order and returns to the finish area in the least time.

SCORE ORIENTEERING

Quick thinking and strategy are major factors in score orienteering. A competitor selects the check-points to find based on point value and location. Point values throughout the course are high or low depending on how hard the markers are to reach. Whoever collects the most points within a designated time is the winner. Points are deducted for returning late to the finish area.

LINE ORIENTEERING

Line orienteering is excellent for training new orienteers. The route is premarked on the map, but checkpoints are not shown. The navagator tries to walk or run the exact map route. While negotiating the course, he looks for checkpoints or control-marker signs. The winner is determined by the time taken to run the course and the accuracy of marking the control points when they are found.

ROUTE ORIENTEERING

This variation is also excellent for beginners. The navigator follows a route that is clearly marked with signs or streamers. While negotiating the course, he records on the map the route being taken. Speed and accuracy of marking the route determine the winner.

NIGHT ORIENTEERING

Competitors in this event carry flashlights and navigate with map and compass. The night course for cross-country orienteering is usually shorter than the day course. Control points are marked with reflective material or dim lights. Open, rolling terrain, which is poor for day courses, is much more challenging at night.

URBAN ORIENTEERING

Urban orienteering is very similar to traditional types, but a compass, topographical map, and navigation skills are not needed. A course can be set up on any installation by using a map of the main post or cantonment area. Soldiers run within this area looking for coded location markers, which are numbered and marked on the map before the start. This eliminates the need for a compass. Soldiers only need a combination map-scorecard, a watch, and a pencil. (Figure 9-7 shows a sample scorecard.)

Urban orienteering adds variety and competition to a unit's PT program and is well suited for an intramural program. It also provides a good cardiovascular workout.

Participants and Rules

Urban orienteering is conducted during daylight hours to ensure safety and make the identification of checkpoint markers easy. Soldiers form two-man teams based on their APFT 2-mile-run times. Team members should have similar running ability. A handicap is given to slower teams. (See Figure 9-8.) At the assembly area, each team gets identical maps that show the

URBAN ORIENTEERING

LOCATION MARKER	POINT VALUE	LOCATION MARKER CODE	LOCATION MARKER	POINT VALUE	LOCATION MARKER CODE
1	10		26	10	
2	10		27	15	
3	15		28	5	
4	10		29	15	
5	15		30	15	
6	10		31	15	
7	25		32	25	
8	15		33	15	
9	25		34	15	
10	15		35	25	
11	15		36	15	
12	25		37	15	
13	15		38	25	
14	15		39	15	
15	25		40	25	
16	15		41	25	
17	25		42	15	
18	10		43	10	
19	10		44	15	
20	15		45	10	
21	10		46	25	
22	5		47	10	
23	15		48	15	
24	10		49	15	
25	10		50	10	

Figure 9-7

location of markers on the course. Location markers are color-coded on the map based on their point value. The markers farthest from the assembly area have the highest point values. The maps are labeled with a location number corresponding to the location marker on the course. A time limit is given, and teams finishing late are penalized. Five points are deducted for each minute a team is late. While on the course, team members must stay together and not separate to get two markers at once. A team that separates is disqualified. Any number of soldiers may participate, the limiting factors being space and the number of points on the course.

Playing the Game

Once the soldiers have been assigned a partner, the orienteering marshal briefs them on the rules and objectives of the game. He gives them their time limitations and a reminder about the overtime penalty. He also gives each team a combination map/scorecard with a two-digit number on it to identify their team. When a team reaches a location marker, it records on the scorecard the letters that correspond to its two-digit number.

Point values of each location marker are also annotated on the scorecard. When the orienteering marshal signals the start of the event, all competitors

HANDICAPS FOR URBAN ORIENTEERING

2-MILE RUN TIME	POINTS	2-MILE RUN TIME	POINTS
12:00 or faster	0	14:31-15:00	60
12:01-12:30	10	15:01-15:30	70
12:31-13:00	20	15:31-16:00	80
13:01-13:30	30	16:01-16:30	90
13:31-14:00	40	16:31-17:00	100
14:01-14:30	50	17:01+	100

Figure 9-8

SAMPLE LOCATION MARKER

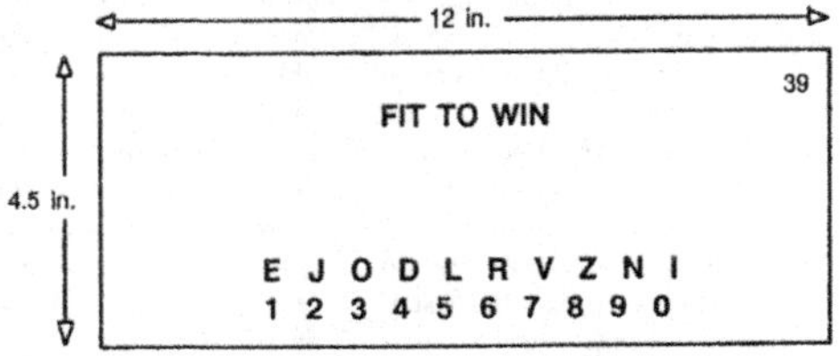

Figure 9-9

leave the assembly area at the same time. One to two hours is the optimal time for conducting the activity. A sample location marker is shown at Figure 9-9.

For this example, team number 54 found the marker. The letters corresponding to 54 are LD, so they place "LD" on line 39 of their scorecard. This line number corresponds to the location's marker number. When the location marker code is deciphered, the team moves on to the next marker of its choice. Each team goes to as many markers as possible within the allotted time. After all teams have found as many location markers as possible and have turned in their map/scorecards, the points are computed by the orienteering marshal to determine the teams' standings. He has the key to all the points and can determine each team's accuracy. Handicap points are then added. Each soldier gets points if his 2-mile-run time is slower than 12 minutes. (See Figure 9-8.) The teams' standings are displayed shortly after the activity ends.

Safety Briefing

The orienteering marshal gives a safety briefing before the event starts. He reminds soldiers to be cautious while running across streets and to emphasize that team members should always stay together.

Set Up and Materials

The course must be well thought out and set up in advance. Setting up requires some man-hours, but the course can be used many times. The major tasks are making and installing location markers and preparing map/scorecard combinations. Once the location marker numbers are marked and color coded on the maps, they are covered with combat acetate to keep them useful for a long time. Combat acetate (also called plastic sheet) can be purchased in the self-service supply center store under stock number 9330-00-618-7214.

The course organizer must decide how many location markers to make and where to put them. He should use creativity to add excitement to the course. Suggestions for locations to put point markers are as follows: at intersections, along roads in the tree line, on building corners, and along creek beds and trails. They should not be too hard to find. To help teams negotiate the course, all maps must be precisely marked to correspond with the placement of the course-location markers.

Unit Olympics

Unit olympics incorporate athletic events that represent all five fitness components.

The unit olympics is a multifaceted event that can be tailored to any unit to provide athletic participation for all soldiers. The objective is to incorporate into a team-level competition athletic. events that represent all five fitness components. The competition can be within a unit or between competing units. When conducted with enthusiasm, it promotes team spirit and provides a good workout. It is a good diversion from the regular PT session.

A unit olympics, if well promoted from the top and well staged by the project NCO or officer, can be a good precursor to an SDT or the EIB test.

TYPES OF EVENTS

The olympics should include events that challenge the soldiers' muscular strength and endurance, aerobic endurance, flexibility, agility, speed, and related sports skills.

Events can be held for both individuals and teams, and they should be designed so that both male and female soldiers can take part. Each soldier should be required to do a minimum number of events. Teams should wear a distinctively marked item such as a T-shirt or arm band. This adds character to the event and sets teams apart from each other. A warm-up should precede and a cool-down should follow the events.

The following are examples of athletic events that could be included in a unit olympics:

Push-Up Derby

This is a timed event using four-member teams. The objective is for the team to do as many correct push-ups as possible within a four-minute time limit. Only one team member does push-ups at a time. The four team members may rotate as often as desired,

Sandbag Relay

This event uses four-man teams for a running relay around a quarter-mile track carrying sandbags. One player from each team lines up at the starting line with a full sandbag *in* each hand. He hands the sandbags off to a teammate when he finishes his part of the race. This continues until the last team player crosses the finish line. Placings are determined by the teams' order of finish.

Team Flexibility

In this event, if teams are numerically equal, all members of each team should participate. If not, as many team members should participate as possible. Each team's anchor person places his foot against a wall or a curb. He stretches his other foot as far away as possible as in doing a split. The next team member puts one foot against the anchor man's extended foot and does a split-stretch. This goes on until all team members are stretched. They cover as much distance as possible keeping in contact with each other. The team that stretches farthest from the start point without a break in their chain is the winner.

Medicine-Ball Throw

This event uses four-member teams. The teams begin by throwing the ball from the same starting line. When it lands, the ball is marked for each team thrower, and the next team player throws from this spot. This is repeated until all the team's players have thrown. The team whose combined throws cover the most distance is the winner.

Job-Related Events

The organizer should use his imagination when planning activities. He may incorporate soldier skills required of an MOS. For instance, he could devise a timed land-navigation event geared toward soldiers with an MOS of 11 C. The team would carry an 81 -mm mortar (tube, tripod, and baseplate) to three different locations, each a mile apart, and set it up in a firing configuration. This type of event is excellent for fine-tuning job skills and is also physically challenging.

OPENING CEREMONY

The commander, ranking person, or ceremony host gives an inspirational speech before the opening ceremonies, welcoming competitors and wishing them good luck. The olympics is officially opened with a torch lighting. This is followed by a short symbolic parade of all the teams. The teams are then put back into formation, and team captains lead motivating chants. The master of ceremonies (MC) announces the sequence of events and rules for each event. The games then begin.

JUDGING AND SCORING

The MC should have one assistant per team who will judge that one team during each event. Assistants give input on events that need a numerical count. The MC monitors the point accumulation of each team. Points are awarded for each event as follows:

- First = 4 points.
- Second = 3 points.
- Third = 2 points.
- Fourth = 1 point.

When two teams tie an event, the points are added together and split equally between them. After the competition ends, the totaled point scores for each team are figured. The first- through fourth-place teams are then recognized.

CHAPTER 10

Developing the Program

Commanders must develop prgrams that train soldiers to maximize their physical performance.

The goal of the Army's physical fitness program is to improve each soldier's physical ability so he can survive and win on the battlefield. Physical fitness includes all aspects of physical performance, not just performance on the APFT. Leaders must understand the principles of exercise, the FITT factors, and know how to apply them in order to develop a sound PT program that will improve all the fitness components. To plan PT successfully, the commander and MFT must know the training management system. (See FM 25-100.)

Commanders should not be satisfied with merely meeting the minimum requirements for physical training which is having all of their soldiers pass the APFT. They must develop programs that train soldiers to maximize their physical performance. Leaders should use incentives. More importantly, they must set the example through their own participation.

The unit PT program is the commander's program. It must reflect his goals and be based on sound, scientific principles. The wise commander also uses his PT program as a basis for building team spirit and for enhancing other training activities. Tough, realistic training is good. However, leaders must be aware of the risks involved with physical training and related activities. They should, therefore, plan wisely to minimize injuries and accidents.

Steps in Planning

STEP 1: ANALYZE THE MISSION

When planning a physical fitness program, the commander must consider the type of unit and its mission. Missions vary as do the physical requirements necessary to complete them. As stated in FM 25-100, "The wartime mission drives training." A careful analysis of the mission, coupled with the commander's intent, yields the mission-essential task list (METL) a unit must perform.

Regardless of the unit's size or mission, reasonable goals are essential. According to FM 25-100, the goals should provide a common direction for all the commander's programs and systems. An example of a goal is as follows because the exceptional physical fitness of the soldier is a critical combat-multiplier in the division, it must be our goal to ensure that our soldiers are capable of roadmarching 12 miles with a 50-pound load in less than three hours.

STEP 2: DEVELOP FITNESS OBJECTIVES

Objectives direct the unit's efforts by prescribing specific actions. The commander, as tactician, and the MFT, as physical fitness advisor, must analyze the METL and equate this to specific fitness objectives. Examples of fitness objectives are the following:

- Improve the unit's overall level of strength by ensuring that all soldiers in the unit can correctly perform at least one repetition with 50 percent of their bodyweight on the overhead press using a barbell.
- Improve the unit's average APFT score through each soldier obtaining a minimum score of 80 points on the push-up and sit-up events and 70 points on the 2-mile run.
- Decrease the number of physical training injuries by 25 percent through properly conducted training.

The commander and MFT identify and prioritize the objectives.

STEP 3: ASSESS THE UNIT

With the training objectives established, the commander and MFT are ready to find the unit's current fitness level and measure it against the desired level.

Giving a diagnostic APFT is one way to find the current level. Another way is to have the soldiers road march a certain distance within a set time while carrying a specified load. Any quantifiable, physically demanding, mission-essential task can be used as an assessment tool. Training records and reports, as well as any previous ARTEP, EDREs, and so forth, can also provide invaluable information.

STEP 4: DETERMINE TRAINING REQUIREMENTS

By possessing the unit's fitness capabilities and comparing them to the standards defined in training objectives, leaders can determine fitness training requirements. When, after extensive training, soldiers cannot reach the desired levels of fitness, training requirements may be too idealistic. Once training requirements are determined, the commander reviews higher headquarters' long- and short-range training plans to identify training events and allocations of resources which will affect near-term planning.

STEP 5: DEVELOP FITNESS TASKS

Fitness tasks provide the framework for accomplishing all training requirements. They identify what has to be done to correct all deficiencies and sustain all proficiencies. Fitness tasks establish priorities, frequencies, and the sequence for training requirements. They must be adjusted for real world constraints before they become a part of the training plan. The essential elements of fitness tasks can be cataloged into four groups:

(1) Collective tasks
(2) Individual tasks
(3) Leader tasks
(4) Resources required for training

Collective tasks. Collective tasks are the training activities performed by the unit. They are keyed to the unit's specific fitness objectives. An example would be to conduct training to develop strength and muscular endurance utilizing a sandbag circuit.

Individual tasks. Individual tasks are activities that an individual soldier must do to accomplish the collective training task. For example, to improve CR endurance the individual soldier must do ability-group running, road marching, Fartlek training, interval training, and calculate/monitor his THR when appropriate.

Leader tasks. Leader tasks are the specific tasks leaders must do in order for collective and individual training to take place. These will involve procuring resources, the setting up of training, education of individual soldiers, and the supervision of the actual training.

Resources. Identifying the necessary equipment, facilities, and training aids during the planning phase gives the trainer ample time to prepare for the training. The early identification and acquisition of resources is necessary to fully implement the training program. The bottom line is that training programs must be developed using resources which are available.

STEP 6: DEVELOP A TRAINING SCHEDULE

The fitness training schedule results from leaders' near-term planning. Leaders must emphasize the development of all the fitness components and follow the principles of exercise and the FITT factors. The training schedule shows the order, intensity, and duration of activities for PT. Figure 10-1 illustrates a typical PT session and its component parts.

There are three distinct steps in planning a unit's daily physical training activities. They are as follows:

1. Determine the minimum frequency of training. Ideally, it should include three cardiorespiratory and three muscular conditioning sessions each weeks. (See the FITT factors in Chapter 1.)
2. Determine the type of activity. This depends on the specific purpose of the training session. (See Figure 10-2.) For more information on this topic, see Chapters 1, 2, and 3.
3. Determine the intensity and time of the selected activity. (See the FITT factors in Chapter 1.)

Each activity period should include a warm-up, a workout that develops cardiorespiratory fitness and/or muscular endurance and strength, and a cool-down. (See Figure 10-1).

At the end of a well-planned and executed PT session, all soldiers should feel that they have been physically stressed. They should also understand the objective of the training session and how it will help them improve their fitness levels.

STEP 7: CONDUCT AND EVALUATE TRAINING

The commander and MFT now begin managing and supervising the day-to-day training. They evaluate how the training is performed by monitoring its intensity, using THR or muscle failure, along with the duration of the daily workout.

The key to evaluating training is to determine if the training being conducted will result in improvements in physical conditioning. If not, the training needs revision. Leaders should

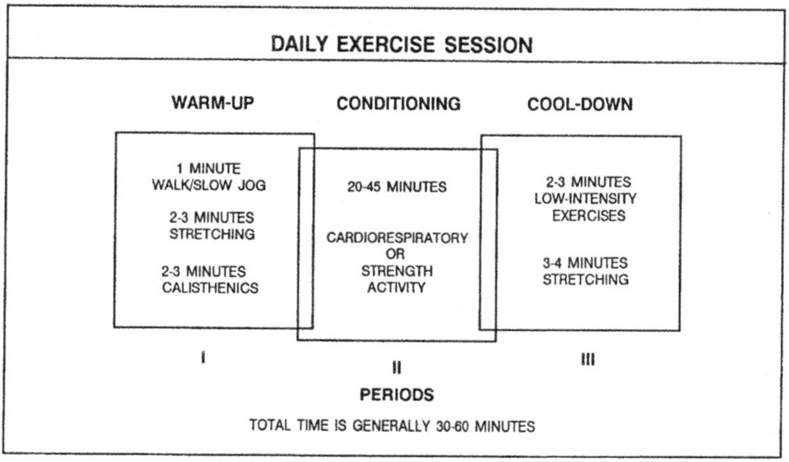

Figure 10-1

ACTIVITY SELECTION GUIDE

PURPOSE	MUSCULAR STRENGTH	MUSCULAR ENDURANCE	CARDIO-RESPIRATORY ENDURANCE	FLEXI-BILITY	BODY COMPO-SITION	SPEED/ AGILITY	COORDI-NATION	TEAM-WORK	SOLDIER SKILLS
Aerobics		X	X	X	X		X		
Bicycling		X	X		X				
Circuits	X	X	X	X	X	X	X	X	X
Competitive Activities						X	X	X	X
Calisthenics		X		X		X	X		
Cross Country Skiing	X	X	X	X	X		X		
Grass/Guerrilla Drills	X	X	X		X		X		
Obstacle Courses	X	X	X		X	X	X	X	X
Partner-Resisted Exercises	X	X					X	X	
Relays		X	X		X	X	X	X	
Rifle Drills	X	X					X		X
Road Marching	X	X	X		X				X
Running		X	X		X				
Stretching				X					
Weight Training	X	X				X	X		

Figure 10-2

not be sidetracked by PT that is all form and little substance. Such training defeats the concept of objective-based training and results in little benefit to soldiers.

Education

Teaching soldiers about physical fitness is vital. It must be an ongoing effort that uses trained experts like MFTs. Soldiers must understand why the program is organized the way it is and what the basic fitness principles are. When they know why they are training in a certain way, they are more likely to wholeheartedly take part. This makes the training more effective.

Education also helps the Army develop its total fitness concept. Total fitness should be reinforced throughout each soldier's career. Classroom instruction in subjects such as principles of exercise, diet and nutrition, tobacco cessation, and stress management should be held at regular intervals. Local "Fit to Win" coordinators (AR 600-63) can help develop classes on such subjects.

Common Errors

There are some common errors in unit programs. The most common error concerns the use of unit runs. When all soldiers must run at the same pace as with a unit run, many do not receive a training effect because they do not reach their training heart rate (THR). The least-fit soldiers of the unit may be at risk because they may be training at heart rates above their THR. Another error is exclusively using activities such as the "daily dozen." These exercises emphasize form over substance and do little to improve fitness.

Yet another error is failing to strike a balance in a PT program between CR endurance training and muscular endurance and strength training. In addition, imbalances often stem from a lack of variety in the program which

Total fitness should be reinforced throughout each soldier's career by classroom instruction.

leads to boredom. The principles of exercise are described in Chapter 1, and their application is shown in the sample program below.

A Sample Program

The following sample program shows a commander's thought processes as he develops a 12-week fitness training program for his unit.

Captain Frank Jones's company has just returned from the field where it completed an ARTEP. Several injuries occurred including a broken foot, resulting from a dropped container, and three low back strains. After evaluating his unit during this ARTEP, CPT Jones concluded that its level of physical fitness was inadequate. He thought this contributed to the injuries and poor performance. The soldiers' flexibility was poor, and there was an apparent lack of prior emphasis on, and training in, good lifting techniques. This, combined with poor flexibility in the low back and hamstrings, may have contributed to the unacceptably high number of low back strains. Captain Jones decided to ask the battalion's MFT to help him develop a good unit program for the company. They went through the following steps.

7-STEP PLANNING PROCESS

ANALYZE THE MISSION

DEVELOP FITNESS OBJECTIVES

ASSESS THE UNIT

DETERMINE TRAINING REQUIREMENTS

DESIGN FITNESS TASKS

DEVELOP A TRAINING SCHEDULE

CONDUCT AND EVALUATE TRAINING

ANALYZE THE MISSION

First, they analyzed the recently completed ARTEP and reviewed the ARTEP manual to find the most physically demanding, mission-oriented tasks the unit performs. The analysis showed that, typically, the company does a tactical road march and then occupies a position. It establishes a perimeter, improves its positions, and selects and prepares alternate positions. One of the most demanding missions while in position requires soldiers to move by hand, for 15 to 30 minutes, equipment weighing up to 95 pounds. If his unit received artillery fire, it would need to be able to move to alternate positions as quickly as possible. This requires much lifting, digging, loading, unloading, and moving of heavy equipment. All of these tasks require good muscular endurance and strength and a reasonable level of cardiorespiratory endurance.

DEVELOP FITNESS OBJECTIVES

Next, CPT Jones reviewed his battalion commander's physical training guidance. It showed that the commander was aware that the unit's tasks require muscular endurance and strength and cardiorespiratory fitness. The guidance and objectives issued are as follows:

a. Units will do PT five days a week (0600-0700) when in garrison. In the field, organized PT will beat the commander's discretion.

Captain Jones determined that the major PT emphasis should be to improve muscular endurance and strength. He based this on his unit's mission, training schedule, available resources, and on his commander's guidance and objectives. With this information and the MIT's recommendations, CPT Jones developed the following fitness objectives.

- Improve the unit's overall level of muscular endurance and strength.
- Improve the unit's overall level of flexibility.
- Improve the unit's average APFT score. Each soldier will score at least 80 points on the push-up and

sit-up events and 70 points on the 2-mile run.

- Improve the unit's road marching capability so that 100 percent of the unit can complete a 12-mile road march with a 35-pound load in at least 3.5 hours.
- Decrease the number of profiles.
- Reduce tobacco use.

ASSESS THE UNIT

The next step CPT Jones accomplished was to assess his unit.

The MFT studied the results of the unit's latest APFT and came up with the following information:

- The average push-up score was 68 points.
- The average sit-up score was 72 points.
- The average number of points scored on the 2-mile run was 74.
- There were six failures, two on the 2-mile run and four on the push-up.

The MFT also recommended that the unit be assessed in the following areas: road march performance, strength, flexibility, substance abuse, and profiled soldiers.

Following the MFT's recommendations, subordinate leaders made the following assessments/determinations:

- Eighty-eight percent of the company finished the 12-mile road march with a 35-pound load in under 3 hours 30 minutes.
- A formation toe-touch test revealed that over half the company could not touch their toes while their knees were extended.
- Thirty percent of the unit uses tobacco.
- Two soldiers are in the overweight program.
- Eight percent of the unit is now on temporary profile, most from back problems.

DETERMINE TRAINING REQUIREMENTS

The next step CPT Jones accomplished was to determine the training requirements.

Training requirements are determined by analyzing the training results and the data obtained from the unit assessment. The next step is to compare this data to the standards identified in the training objectives. When performance is less than the established standard, the problem must be addressed and corrected.

Captain Jones established the following training requirements.

Units will do flexibility exercises during the warm-up and cool-down phase of every PT session. During the cool-down, emphasis on will be placed on developing flexibility in the low back, hamstrings, and hip extensor muscle groups.

Each soldier will do 8 to 12 repetitions of bent-leg, sandbag dead-lifts at least two times a week to develop strength. The section leader will supervise lifts.

Each soldier will do heavy resistance/weight training for all the muscle groups of the body two to three times a week.

Each soldier will perform timed sets of push-ups and sit-ups.

Each soldier will train at least 20 to 30 minutes at THR two to three times a week.

Road marches will be conducted at least once every other week.

Tobacco cessation classes will be established to reduce the number of tobacco users.

DESIGN FITNESS TASKS

Once all training requirements are identified, the next step is to use them to design fitness tasks which relate to the fitness objectives. In developing the fitness tasks, CPT Jones must address collective, individual, and leader tasks as well as resources required.

Fitness tasks provide the framework for accomplishing the training requirements. By accurately listing the fitness tasks that must be done and the resources required to do them, the subsequent step of developing a training schedule is greatly facilitated.

An example of designing fitness tasks is provided in Figure 10-3 by using the activities which might occur during one week of physical training.

The collective tasks for the unit are to perform the following: develop muscular endurance and strength, improve CR endurance, and improve flexibility.

The individual tasks all soldiers must perform during the week are as follows. For developing strength and muscular endurance, they must perform appropriate strength circuit exercises, PREs, sandbag circuits, to include performing bent-leg dead lifts exercises, and training for push-up/sit-up improvement. To improve cardiorespiratory endurance, they must do ability-group runs, interval training, road marching, and they must calculate their THR and monitor THR when appropriate. To improve their flexibility, they must do stretching exercises during their daily warm-up and cool-down.

The leader's tasks are to organize and supervise all strength- and muscle endurance-training sessions and CR training sessions so as to best meet all related fitness objectives. Similarly, the leader must organize and supervise all warm-up and cool-down sessions to best meet the fitness objectives for the development and maintenance of flexibility.

To provide specific examples of leaders tasks in the area of training for strength and muscle endurance, the leader will ensure the following:

- Each strength- and/or muscle endurance-training session works all the major muscle groups of the body.
- High priority is given to training those muscles and muscle groups used in mission-essential tasks.
- Areas where weaknesses exist, with respect to strength/muscle

FITNESS TASKS FOR ONE WEEK OF PHYSICAL TRAINING

COLLECTIVE	INDIVIDUAL	LEADER	RESOURCES
Improve strength and muscle endurance	Do STR CIR EX*, PRE, SNDBG CIR, SU-PU IMP	Organize & supervise STR CIR EX, PRE, & SNDBG CIR	STR RM, Gym, Sandbags, PT Field
Improve CR endurance	Do AGR, CAL/MON THR, road march, do intervals (4x440) IND AB	Organize & supervise CR workouts, CAL/MON THR, MON work/relief ratio for intervals	Track, Running Trails
Improve flexibility	Do stretching exercises	Organize & supervise activity	Gym

* A list of abbreviations appears at the end of Figure 10-4.

Figure 10-3

endurance, are targeted in all workouts.

- Problem areas related to APFT performance are addressed in appropriate workouts.
- The duration of each strength training session is 20-40 minutes.
- Soldiers train to muscle failure.
- All the principles of exercise, to include regularity, overload, recovery, progression, specificity, balance are used.

In a similar manner, the leader would ensure that the guidelines and principles outlined in this and earlier chapters are used to organize training sessions for improving CR endurance and flexibility.

The resources needed for the one-week period are as follows: a strength room, a gym, a PT field, a running track and/or running trails, and sandbags.

DEVELOP A TRAINING SCHEDULE

The next step was to develop a fitness training schedule (shown at Figure 10-4). It lists the daily activities and their intensity and duration.

12-WEEK TRAINING PLAN

JULY

MONDAY	TUESDAY	WEDNESDAY	THURSDAY	FRIDAY
START ASSESSMENT* ACT: AGR** INT: 70% HRR*** DUR: 20 MIN	FINISH ASSESSMENT ACT: PLT 1 & 2 STR CIR; PLT 3 & 4 SNDBG CIR/PU-SU IMP INT: MF/MF DUR: 30/4 MIN	ACT: AGR/LINE SOCCER INT: 70% HRR/NA DUR: 20/30 MIN	ACT: PRE/PU-SU IMP INT: MF DUR: 35/4 MIN	ACT: ROAD MARCH, 5 MLE W 35 LBS IN 90 MIN
ACT: PLT 1 & 2 WT STR CIR; PLT 3 & 4 SNDBG CIR/PU-SU IMP INT: MF DUR : 30-35/4 MIN	ACT: AGR/ PAR COURSE INT: 70% HRR DUR: 20/15-20 MIN	FLIP - FLOP MONDAY's WORKOUT	ACT: AGR/GDR INT: 70% HRR DUR: 20/15-20 MIN	ACT: PRE/PU-SU IMP INT: MF DUR: 35/4 MIN
ACT: FIXED CIR I INT: 70% HRR DUR: 30-40 MIN	ACT: PRE/PU-SU IMP INT: MF DUR: 40/5 MIN	ACT: AGR/GDR INT: 70% HRR DUR: 22/20 MIN	ACT: SNDBG CIR/PU-SU IMP INT: MF DUR: 35-40/5 MIN	ACT: ROAD MARCH, 5 MLE W 35 LBS IN 90 MIN
ACT: PLT 1 & 2 STR CIR; PLT 3 & 4 SNDBG CIR/PU-SU IMP INT: MF DUR: 30-40/5 MIN	ACT: AGR INT: 70% HRR DUR: 25 MIN	FLIP - FLOP MONDAY'S WORKOUT	ACT: AGR INT: 70% HRR DUR: 25 MIN	ACT: OBS CRS/ PRE/PU-SU IMP INT: MF DUR: 25/20/5 MIN

* Initially, assessments must be made of each soldier's level of physical fitness. Particularly important is assessing a soldier's strength and muscular endurance by determining his 8-12 RM for each resistance exercise he will be doing. As mentioned in the Phases of Conditioning section in Chapter 3, this will take two weeks and should be planned for accordingly. The other components of fitness should also be addressed as the need arises.

** A list of abbreviations and acronyms appears at the end of this training plan.

*** Those soldiers with a fairly good level of CR fitness (that is, the average soldier) should exercise at about 70 percent HRR. Those with very high levels of CR fitness may benefit most from training at around 80 to 85 percent HRR during a CR training workout.

Figure 10-4

12-WEEK TRAINING PLAN

AUGUST

MONDAY	TUESDAY	WEDNESDAY	THURSDAY	FRIDAY
ACT: AGR INT: 70% HRR DUR: 27 MIN	ACT: PLT 1 & 2 SNDBG CIR; PLT 3 & 4 PRE/PU-SU IMP INT: MF DUR: 40/6	ACT: INTERVALS INT: 8 x 440 IND AB DUR: 45 MIN	FLIP - FLOP TUESDAY'S WORKOUT	ACT: ROAD MARCH, 7.5 MLE W 35 LBS IN 2.5 HOURS
ACT: PLT 1 & 2 SNDBG CIR; PLT 3 & 4 STR CIR/PU SU IMP INT: MF DUR: 40/6 MIN	ACT: LAST MAN-UP RUN IN AG/ PAR CRS INT: 70-80% HRR*/70%HRR DUR: 30/20 MIN	FLIP - FLOP MONDAY'S WORKOUT	ACT: AGR/FIT-NESS RELAYS INT: 70% HRR/NA DUR: 30/20 MIN	ACT: PLT 1 & 2 SNDBG CIR; PLT 3 & 4 PRE/ PU-SU IMP INT: MF DUR: 40/6 MIN
IN FIELD: PLAN FOR	IN FIELD PLAN FOR	IN FIELD: PLAN FOR	IN FIELD: PLAN FOR	IN FIELD: PLAN FOR
ACT: LAST MAN-UP RUN IN AG INT: 70-90% HRR* DUR: 32 MIN	ACT: PRE/PU-SU IMP INT: MF DUR: 40/7 MIN	ACT: FARTLEK IN AG INT: 60-90% HRR* DUR: 32 MIN	ACT: PRE/PU-SU IMP INT: MF DUR: 40/8 MIN	ACT: ROAD MARCH, 10 MLE W 35 LBS IN 3.5 HOURS
ACT: PRE/PU-SU IMP INT: MF DUR: 35/10 MIN	ACT: INTERVALS INT: 8 x 440 IND AB DUR: 45 MIN	ACT: PU-SU, PULL-UP IMP INT: MF DUR: 45 MIN	ACT: FARTLEK IN AG INT: 60-90% HRR* DUR: 35 MIN	ACT: PU-SU, PULL-UP IMP INT: MF DUR: 45 MIN LC: APFT FOR GRADERS

* During the Last-Man-Up and Fartlek running, the heart rate will vary depending on whether it is taken during the slower or the faster portion of the run. The smaller and larger numbers provided for percent HRR should set the lower and upper limits, respectively, for a soldier's heart rate during this type of training. During interval running, the soldier should concern himself with running at the appropriate pace; he should not monitor THR during interval work.

Figure 10-4 (continued)

12-WEEK TRAINING PLAN

SEPTEMBER

MONDAY	TUESDAY	WEDNESDAY	THURSDAY	FRIDAY
ACT: DEVEL-OPMENTAL STRETCHING INT: SLIGHT TENSION, NOT PAIN DUR: 20-30 MIN	ACT: APFT DUR: NA	ACT: UNIT OLYMPICS, PART I DUR: NA	ACT: UNIT OLYMPICS, PART II DUR: NA	ACT: APFT & OLYMPIC AWARDS CEREMONY/ UNIT RUN INT: NA/CD DUR: NA/30-40 MIN
ACT: PLT 1 & 2 STR CIR; PLT 3, & 4 PRE INT: MF DUR: 40 MIN	ACT: PLT 1 & 2 AGR; PLT 3 & 4 EX TO MUSIC INT: 70% HRR DUR: 35/45 MIN	FLIP - FLOP MONDAY'S WORKOUT	ACT: ROAD MARCH, 10 MLE W 35 LBS IN 3 HOURS	ACT: PLT 1 & 2 SNDBG CIR; PLT 3 & 4 PRE INT; MF DUR: 40 MIN
ACT: AGR INT: 70% HRR DUR: 35 MIN	ACT: PRE/PU-SU IMP INT: MF DUR: 40/8 MIN	ACT: FIXED CIRCUIT/RELAYS INT: 70% HRR/NA DUR: 20/20 MIN	ACT: UPPER BODY PRE/PU-SU IMP INT: MF DUR: 30/8 MIN	ACT: ROAD MARCH, 12 MLE W 35 LBS IN UNDER 4 HOURS
ACT: PLT 1 & 2 LOG DRILLS; PLT 3 & 4 SNDBG CIR/PU-SU IMP INT: MF DUR: 30/8 MIN	ACT: PLT 1 & 2 FIXED CIR; PLT 3 & 4 AQUATICS INT: 70% HRR/NA DUR: 30 MIN	FLIP - FLOP MONDAY'S WORKOUT	FLIP - FLOP TUESDAY'S WORKOUT	ACT: PRE/PU-SU IMP INT: MF DUR: 35/8 MIN

NOTES

1. Push-ups and sit-ups are done as part of each strength workout. In the above sessions, they have been placed near the end of the workout. However, they can occasionally be done before the strength workout for variety. An example of a beginning PU-SU improvement workout lasting about three minutes follows:

a. Perform one timed set of push-ups for 50 seconds. Follow this immediately with one 50-second, timed set of sit-ups. For all timed sets, each soldier must perform as many repetitions of the exercise as he can during the alloted time period.

b. Perform a second set of push-ups for 40 seconds. Follow this immediately with a timed set of sit-ups of equal duration.

As the soldier adapts to this, the difficulty of the session can be increased by adding more timed sets and/or by decreasing the rest interval between like or unlike sets of exercises. For example, the rest period between timed sets of push-ups and sit-ups can be decreased. Also, all of the timed sets for push-ups may be done back-to-back (as can the sit-ups), the rest interval between these timed sets of push-ups can be progressively reduced to make the workout more demanding. Many more options exist for increasing the difficulty of, and adding variety to, these sessions.

2. Activities are planned for the FTX; duration is determined on site.
3. The unit's olympic events include the following:
   a. Ammo-box shuttle-run (fastest time by section).
   b. Biceps, barbell curl (most reps with 60 lbs., total by section).
   c. Leg press (most weight lifted by section).
   d. Standing-toe touch (most soldiers touching toes by section/must hold five seconds).
   e. Highest APFT score by section.

Figure 10-4 (continued)

ABBREVIATIONS AND ACRONYMS

ACT	activity
AG	ability groups
AGR	ability group run
ANA ACT	anaerobic activity
AOTR	assessment of training requirements
CAL/MON	calculate/monitor
CD	commander's decision
CFA	competitive fitness activities
CIR	circuit
CR	cardiorespiratory training
DL	dead-lift (bent-leg)
EX	exercise
GDR	grass drills
GUD	guerilla drills
HRR	heart rate reserve
IMP	improvement
IND AB	at the individual's ability
INT	intensity
LBS	pounds
LC	leader's class
MIN	minute(s)
MF	muscle failure (due to fatigue)
MLE	mile(s)
MS/E	muscle strength/endurance
NA	not applicable
OBS CRS	obstacle course
PLT	platoon
PRE	partner-resisted exercise(s)
PT FLD	physical training field
PU-SU IMP	push-up, sit-up improvement
R	run
SNDBG	sandbag
STR CIR EX	strength circuit exercise
STR RM	strength room
THR	training heart rate
TNG	training
W	with
WT STR CIR	strength circuit with weights
% HRR	percent of heart rate reserve
2-MR	2-mile run

Figure 10-4 (continued)

CONDUCT AND EVALUATE TRAINING

Conducting and evaluating training is the final phase of the training process. This phase includes the evaluation of performance, assessment of capabilities, and feedback portions of the training management cycle. These portions of the cycle must be simultaneous and continuous. To be effective, the evaluation process must address why weaknesses exist, and it must identify corrective actions to be taken. Evaluations should address the following:

- Assessment of proficiency in mission-essential tasks.
- Status of training goals and objectives.
- Status of training in critical individual and collective tasks.
- Shortfalls in training.
- Recommendations for next training cycle (key in on correcting weaknesses).
- Results of educational programs.

Using the Principles of Exercise

As CPT Jones developed his program, he made sure he used the seven principles of exercise. He justified his program as follows:

- Balance. This program is balanced because all the fitness components are addressed. The emphasis is on building muscular endurance and strength in the skeletal muscular system because of the many lifting tasks the unit must do. The program also trains cardiorespiratory endurance and flexibility, and warm-up and cool-down periods are included in every workout.
- Specificity. The unit's fitness goals are met. The sand-bag lifting and weight training programs help develop muscular endurance and strength. The movements should, when possible, stress muscle groups used in their job-related lifting tasks. Developmental stretching should help reduce work-related back injuries. The different types of training in running will help ensure that soldiers reach a satisfactory level of CR fitness and help each soldier score at least 70 points on the APFT's 2-mile run. Soldiers do push-ups and sit-ups at least two or three times a week to improve the unit's performance in these events. The competitive fitness activities will help foster teamwork and cohesion, both of which are essential to each section's functions.
- Overload. Soldiers reach overload in the weight circuit by doing each exercise with an 8- to 12-RM lift for a set time and/or until they reach temporary muscle failure. For the cardiorespiratory workout, THR is calculated initially using 70 percent of the HRR. They do push-ups and sit-ups in multiple, timed sets with short recovery periods to ensure that muscle failure is reached. They also do PREs to muscle failure.
- Progression. To help soldiers reach adequate overload as they improve, the program is made gradually more difficult. Soldiers progress in their CR workout by increasing the time they spend at THR up to 30 to 45 minutes per session and by maintaining THR. They progress on the weight training circuit individually. When a soldier can do an exercise for a set time without reaching muscle failure, the weight is increased so that the soldier reaches muscle failure between the 8th and 12th repetition again. Progression in push-ups and sit-ups involves slowly increasing the duration of the work intervals.
- Variety. There are many different activities for variety. For strength and muscular endurance training the soldiers use weight circuits, sandbag circuits, and PREs. Ability group runs, intervals, Par courses,

Fartlek running, and guerrilla drills are all used for CR training. Varied stretching techniques, including static, partner-assisted, and contract-relax, are used for developmental stretching.

- Regularity. Each component of fitness is worked regularly. Soldiers will spend at least two to three days a week working each of the major fitness components. They will also do push-ups and sit-ups regularly to help reach their peak performance on the APFT.
- Recovery. The muscular and cardiorespiratory systems are stressed in alternate workouts. This allows one system to recover on the day the other is working hard.

Conclusion

CPT Jones's step-by-step process of developing a sound PT program for his unit is an example of what each commander should do in developing his own unit program.

Good physical training takes no more time to plan and execute than does poor training. When commanders use a systematic approach to develop training, the planning process bears sound results and the training will succeed.

CHAPTER 11

Physical Training During Initial Entry Training

Soldiers report to initial entry training (IET) ranging widely in their levels of physical fitness. Because of this, there are special considerations when designing a physical training program for IET soldiers. Physical training involves safely training and challenging all soldiers while improving their fitness level to meet required standards. The regulations which govern the conduct of physical training in IET and explain the graduation requirements are TRADOC Reg. 350-6 and AR 350-15.

The mission of physical training in IET is twofold: to safely train soldiers to meet the graduation requirements of each course and to prepare soldiers to meet the physical demands of their future assignments.

Program Development

All physical training programs in IET must do the following: 1) progressively condition and toughen soldiers for military duties; 2) develop soldiers' self-confidence, discipline, and team spirit; 3) develop healthy life-styles through education; and, 4) improve physical fitness to the highest levels possible in all five components of physical fitness (cardiorespiratory endurance, muscular strength, muscular endurance, flexibility, and body composition).

Because each IET school is somewhat different, commanders must examine the graduation requirements for the course and establish appropriate fitness objectives. They can then design a program that attains these objectives. The seven principles of exercise outlined in Chapter 1 are universal, and they apply to all PT programs including those in IET. Commanders of initial entry training should look beyond the graduation requirements of their own training course to ensure that their soldiers are prepared for the physical challenges of their future assignments. This means developing safe training programs which will produce the maximum physical improvement possible.

MFTs are skilled at assessing soldiers' capabilities. They use the five components of physical fitness in designing programs to reach the training objectives established by the commander. They also know how to conduct exercise programs that are effective and safe. MFTs are not, however, trained to diagnose or treat injuries.

The commander's latitude in program development varies with the length and type of the IET course. For example, commanders of basic combat training (BCT) may do a standard PT program at one installation, while AIT commanders may design their own programs. Regardless of the type of course, all leaders must strive to train their soldiers to attain the highest level of physical fitness possible. This means using the established principles of exercise to develop a safe physical training program.

Safety Considerations

Overuse injuries are common in IET. However, they can be avoided by carefully following the exercise principles of "recovery" and "progression."

Research suggests that soldiers are more prone to injuries of the lower extremities after the third week of IET. High-impact activities, such as road marching and running on hard surfaces, should be carefully monitored during at this time. During this period, fixed circuits and other activities that develop CR fitness are good, low-impact alternatives.

Properly fitted, high-quality running shoes are important, especially when PT sessions require running on hard surfaces. Court shoes, like basketball or tennis shoes, are not designed to absorb the repetitive shock of running. Activities such as running obstacle courses and road marching require combat boots to protect and support the feet and ankles. Naturally, common sense dictates a reasonable break-in period for new combat boots, especially before long marches.

Examples of recommended PT sessions and low-risk exercises are in Chapter 7. Specific health and safety considerations are in TRADOC Reg. 350-6, paragraph 4-2.

Road Marching

One road march should be conducted weekly with the difficulty of the marches progressing gradually throughout IET.

In the first two weeks of IET, soldiers can be expected to road march up to 5 kilometers with light loads. Loads should be restricted to the standard LCE, kevlar helmet, and weapon. Bones, ligaments, and tendons respond slowly to training and may be injured if the load and/or duration are increased too quickly.

After the initial adaptations in the early weeks of IET, soldiers can be expected to carry progressively heavier loads including a rucksack. By he start of the fourth week, they should be accustomed to marching in boots, and their feet should be less prone to blistering. By the sixth week, the load may be increased to 40 pounds including personal clothing and equipment. At no time during IET or one-station unit training (OSUT) should loads exceed 40 pounds.

A sample regimen for road marches during IET is at Figure 11-1.

SAMPLE ROAD MARCH PROGRAM

WEEK	DISTANCE	EQUIPMENT	REMARKS
1	5 km	LCE, kevlar helmet, weapon	
2	5 km	SAME AS WEEK 1	
3	7 km	LCE, kevlar helmet, weapon, and 10-pound rucksack	With all equipment, the total load is 30 pounds.
4	7 km	SAME AS WEEK 3	SAME TOTAL LOAD AS WEEK 3
5	7 km	SAME AS WEEK 3	SAME TOTAL LOAD AS WEEK 3
6	10 km	LCE, hevlar helmet, weapon, and 20-pound rucksack	With all equipment, the total load is 40 pounds.
7	10 km	SAME AS WEEK 6	SAME TOTAL LOAD AS WEEK 6
8	10 km	SAME AS WEEK 6	SAME TOTAL LOAD AS WEEK 6

Note: The total load carried (to include the LCE, kevlar helmet, weapon, and ruck load) should not exceed that shown in the remarks column. If the road marches are to or from non-tactical training, they need not be tactical road marches.

Figure 11-1

CHAPTER 12

Environmental Considerations

In today's Army, soldiers may deploy anywhere in the world. They may go into the tropical heat of Central America, the deserts of the Middle East, the frozen tundra of Alaska, or the rolling hills of Western Europe. Each environment presents unique problems concerning soldiers' physical performance. Furthermore, physical exertion in extreme environments can be life-threatening. While recognizing such problems is important, preventing them is even more important. This requires an understanding of the environmental factors which affect physical performance and how the body responds to those factors.

Temperature Regulation

The body constantly produces heat, especially during exercise. To maintain a constant normal temperature, it must pass this heat on to the environment. Life-threatening circumstances can develop if the body becomes too hot or too cold. Body temperature must be maintained within fairly narrow limits, usually between 74 and 110 degrees Fahrenheit. However, hypothermia and heat injuries can occur within much narrower limits. Therefore, extreme temperatures can have a devastating effect on the body's ability to control its temperature.

Overheating is a serious threat to health and physical performance. During exercise, the body can produce heat at a rate 10 to 20 times greater than during rest. To survive, it must get rid of the excess heat.

The four ways in which the body can gain or lose heat are the following:

- Conduction-the transfre of heat from a warm object to a cool one that is touching it. (Warming boots by putting them on is an example.)
- Convection-the transfer of heat by circulation or movement of air. (Using a fan on a hot day is an example.)
- Radiation-the transfer of heat by electromagnetic waves. (Sitting under a heat lamp is an example.)
- Evaporation- the transfer of heat by changing a liquid into a gas. (Evaporating sweat cooling the skin is an example.)

Heat moves from warm to cool areas. During exercise, when the body is extremely warm, heat can be lost by a combination of the four methods. Sweating, however, is the body's most important means for heat loss, especially during exercise. Any condition that slows or blocks the transfer of heat from the body by evaporation causes heat storage which results in an increase in body temperature.

The degree to which evaporative cooling occurs is also directly related to the air's relative humidity (a measure of the amount of water vapor in the air). When the relative humidity is 100 percent, the air is completely saturated at its temperature. No more water can evaporate into the surrounding air. As a result, sweat does not evaporate, no cooling effect takes place, and the body temperature increases. This causes even more sweating. During exercise in the heat, sweat rates of up to two quarts per hour are not uncommon.

If the lost fluids are not replaced, dehydration can occur. This condition, in turn, can result in severe heat injuries.

Thus, in hot, humid conditions when a soldier's sweat cannot evaporate, there is no cooling effect through the process of evaporation. High relative humidities combined with high temperatures can cause serious problems. Weather of this type occurs in the tropics and equatorial regions such as Central America and southern Asia. These are places where soldiers have been or could be deployed.

Heat Injuries and Symptoms

The following are common types of heat injuries and their symptoms.

- Heat cramps-muscles cramps of the abdomen, legs, or arms.
- Heat exhaustion-headache, excessive sweating, dizziness, nausea, clammy skin.
- Heat stroke-hot, dry skin, cessation of sweating, rapid pulse, mental confusion, unconsciousness.

Adapting to differing environmental conditions is called acclimatization.

To prevent heat injuries while exercising, trainers must adjust the intensity to fit the temperature and humidity. They must ensure that soldiers drink enough water before and during the exercise session. Body weight is a good gauge of hydration. If rapid weight loss occurs, dehydration should be suspected. Plain water is the best replacement fluid to use. Highly concentrated liquids such as soft drinks and those with a high sugar content may hurt the soldier's performance because they slow the absorption of water from the stomach.

To prevent heat injuries, the following hydration guidelines should be used:

- Type of drink: cool water (45 to 55 degrees F).
- Before the activity: drink 13 to 20 ounces at least 30 minutes before.
- During the activity: drink 3 to 6 ounces at 15 to 30 minute intervals.
- After the activity: drink to satisfy thirst, then drink a little more.

Acclimatization to Hot, Humid Environments

Adapting to differing environmental conditions is called acclimatization. Soldiers who are newly introduced to a hot, humid climate and are moderately active in it can acclimatize in 8 to 14 days. Soldiers who are sedentary take much longer. Until they are acclimatized, soldiers are much more likely to develop heat injuries.

A soldier's ability to perform effectively in hot, humid conditions depends on both his acclimatization and level of fitness. The degree of heat stress directly depends on the relative workload. When two soldiers do the same task, the heat stress is less for the soldier who is in better physical condition, and his performance is likely to be better. Therefore, it is important to maintain high levels of fitness.

Increased temperatures and humidity cause increased heart rates. Consequently, it takes much less effort to elevate the heart rate into the training zone, but the training effect is the same. These facts underscore the need to use combat-development running and to monitor heart rates when running, especially in hot, humid conditions.

Some important changes occur as a result of acclimatization to a hot climate. The following physical adaptations help the body cope with a hot environment

- Sweating occurs at a lower body temperature.
- Sweat production is increased.
- Blood volume is increased.
- Heart rate is less at any given work rate.

Exercising in Cold Environments

Contrary to popular belief, there are few real dangers in exercising at temperatures well below freezing. Since the body produces large amounts of heat during exercise, it has little trouble maintaining a normal temperature. There is no danger of freezing the lungs. However, without proper precautions, hypothermia, frostbite, and dehydration can occur.

HYPOTHERMIA

If the body's core temperature drops below normal, its ability to regulate its temperature can become impaired or lost. This condition is called hypothermia. It develops because the body cannot produce heat as fast as it is losing it. This can lead to death. The chance of a soldier becoming hypothermic is a major threat any time he is exposed to the cold.

Some symptoms of hypothermia are shivering, loss of judgment, slurred speech, drowsiness, and muscle weakness.

Hypothermia develops when the body cannot produce heat as fast as it is losing it.

During exercise in the cold, people usually produce enough heat to maintain normal body temperature. As they get fatigued, however, they slow down and their bodies produce less heat. Also, people often overdress for exercise in the cold. This makes the body sweat. The sweat dampens the clothing next to the skin making it a good conductor of heat. The combination of decreased heat production and increased heat loss can cause a rapid onset of hypothermia.

Some guidelines for dressing for cold weather exercise are shown in Figure 12-1.

GUIDELINES FOR DRESSING FOR EXERCISE IN THE COLD

Clothing **for cold weather should protect, insulate, and ventilate.**

- **Protect by covering as large an area of the body as possible.**
- **Insulation will occur by trapping air which has been warmed by the body and holding it near the skin.**
- **Ventilate by allowing a two-way exchange of air through the various layers of clothing.**

Clothing should leave your body slightly cool rather than hot.

Clothing should also be loose enough to allow movement.

Clothing soaked with perspiration should be removed if reasonably possible.

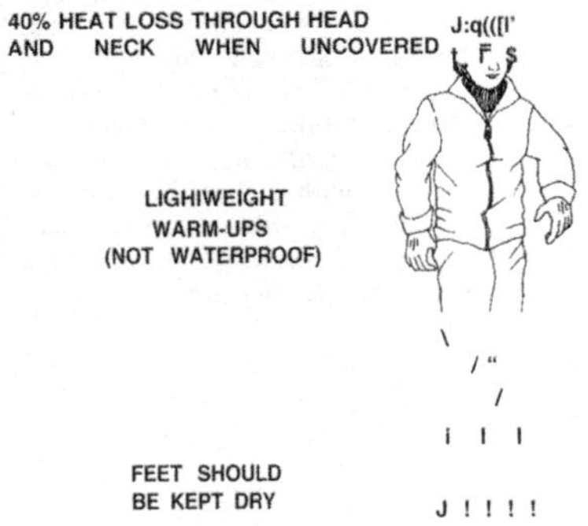

Figure 12-1

FROSTBITE

Frostbite is the freezing of body tissue. It commonly occurs in body parts located away from the core and exposed to the cold such as the nose, ears, feet, hands, and skin. Severe cases of frostbite may require amputation.

Factors which lead to frostbite are cold temperatures combined with windy conditions. The wind has a great cooling effect because it causes rapid convective heat transfer from the body. For a given temperature, the higher the wind speed, the greater the cooling effect. Figure 12-2 shows how the wind can affect cooling by providing information on windchill factors.

A person's movement through the air creates an effect similar to that caused by wind. Riding a bicycle at 15 mph is the same as standing in a 15-mph wind. If, in addition, there is a 5-mph headwind, the overall effect is equivalent to a 20-mph wind. Therefore, an exercising soldier must be very cautious to avoid getting frostbite. Covering exposed parts of the body will substantially reduce the risks.

DEHYDRATION

Dehydration can result from losing body fluids faster than they are replaced. Cold environments are often dry, and water may be limited. As a result, soldiers may in time become dehydrated. While operating in extremely cold climates, trainers should check the body weights of the soldiers regularly and encourage them to drink liquids whenever possible.

WINDCHILL FACTOR

Cooling Power of Wind on Exposed Flesh Expressed as an Equivalent Temperature (under calm conditions)

Estimated wind speed (in mph)	Actual Thermometer Reading (°F)											
	50	40	30	20	10	0	−10	−20	−30	−40	−50	−60
	EQUIVALENT CHILL TEMPERATURE (°F)											
Calm	50	40	30	20	10	0	−10	−20	−30	−40	−50	−60
5	48	37	27	16	6	−5	−15	−26	−36	−47	−57	−68
10	40	28	16	4	−9	−24	−33	−46	−58	−70	−83	−95
15	36	22	9	−5	−18	−32	−45	−58	−72	−85	−99	−112
20	32	18	4	−10	−25	−39	−53	−67	−82	−96	−110	−124
25	30	16	0	−15	−29	−44	−59	−74	−88	−104	−118	−133
30	28	13	−2	−18	−33	−48	−63	−79	−94	−109	−125	−140
35	27	11	−4	−21	−35	−51	−67	−82	−98	−113	−129	−145
40	26	10	−6	−21	−37	−53	−69	−85	−100	−116	−132	−148
(wind speed greater than 40mph have little additional effect)	LITTLE DANGER In 0.5 hr with dry skin Maximum danger of false sense of security				INCREASING DANGER Danger from freezing of exposed flesh within one minute			GREAT DANGER Flesh may freeze within 30 seconds				
	Trenchfoot and immersion foot may occur at any point on this chart.											

INSTRUCTIONS

MEASURE local temperature and wind speed if possible. If not, ESTIMATE. Enter table at closest 5°F interval along the top and with appropriate wind speed along left side. Intersection gives approximate equivalent chill temperature. That is, the temperature that would cause the same rate of cooling under calm conditions. Note that regardless of cooling rate, you do not cool below the actual air temperature unless wet.

Figure 12-2

Acclimatization to High Altitudes

Elevations below 5,000 feet have little noticeable effect on healthy people. However, at higher elevations the atmospheric pressure is reduced, and the body tissues get less oxygen. This means that soldiers cannot work or exercise as well at high altitudes. The limiting effects of high elevation are often most pronounced in older soldiers and persons with low levels of fitness.

Due to acclimatization, the longer a soldier remains at high altitude, the better his performance becomes. Generally, however, he will not perform as well as at sea level and should not be expected to. For normal activities, the time required to acclimatize depends largely on the altitude. In order to insure that soldiers who are newly assigned to altitudes above 5,000 feet are not at a disadvantage, it is recommended that 30 days of acclimatization, including regular physical activity, be permitted before they are administered a record APFT.

Before acclimatization is complete, people at high altitudes may suffer acute mountain sickness. This includes such symptoms as headache, rapid pulse, nausea, loss of appetite, and an inability to sleep. The primary treatment is further acclimatization or returning to a lower altitude.

Once soldiers are acclimatized to altitudes above 5,000 feet, deacclimatization will occur if they spend 14 or more days at lower altitudes. For this reason, soldiers should be permitted twice the length of their absence, not to exceed 30 days, to reacclimatize before being required to take a record APFT. A period of 30 days is adequate for any given reacclimatization.

Air Pollution and Exercise

Pollutants are substances in the environment which lower the environment's quality. Originally, air pollutants were thought to be only byproducts of the industrial revolution. However, many pollutants are produced naturally. For example, volcanoes emit sulfur oxides and ash, and lightning produces ozone.

There are two classifications of air pollutants - primary and secondary. Primary pollutants are produced directly by industrial sources. These include carbon monoxide (CO), sulfur oxides (SO), hydrocarbons, and particulate (ash). Secondary pollutants are created by the primary pollutant's interaction with the environment. Examples of these include ozone **(03)**, aldehydes, and sulfates. Smog is a combination of primary and secondary pollutants.

Some pollutants have negative effects on the body. For example, carbon monoxide binds to hemoglobin in the red blood cells and reduces the amount of oxygen carried in the blood. Ozone and the oxides irritate the air passageways in the lungs, while other pollutants irritate the eyes.

Pollutants can irritate the respiratory tract and make the person less able to perform aerobically.

When exercisers in high-pollution areas breathe through the mouth, the nasal mucosa's ability to remove impurities is bypassed, and many pollutants can be inhaled. This irritates the respiratory tract and makes the person less able to perform aerobically.

The following are some ways to deal with air pollution while exercising:

- Avoid exposure to pollutants before and during exercise, if possible.
- In areas of high ozone concentration, train early in the day and after dark.
- Avoid exercising near heavily traveled streets and highways during rush hours.
- Consult your supporting preventive-medicine activity for advice in identifying or defining training restrictions during periods of heavy air pollution.

CHAPTER 13

Injuries

Most injuries can be prevented by designing a well-balanced PT program.

Injuries are not an uncommon occurrence during intense physical training. It is, nonetheless, a primary responsibility of all leaders to minimize the risk of injury to soldiers. Safety is always a major concern.

Most injuries can be prevented by designing a well-balanced PT program that does not overstress any body parts, allows enough time for recovery, and includes a warm-up and cool-down. Using strengthening exercises and soft, level surfaces for stretching and running also helps prevent injuries. If, however, injuries do occur, they should be recognized and properly treated in a timely fashion. If a soldier suspects that he is injured, he should stop what he is doing, report the injury, and seek medical help.

Many common injuries are caused by overuse, that is, soldiers often exercise too much and too often and with too rapid an increase in the workload. Most overuse injuries can be treated with rest, ice, compression, and elevation (RICE). Following any required first aid, health-care personnel should evaluate the injured soldier.

Typical Injuries Associated with Physical Training

Common injuries associated with exercise are the following:

- Abrasion (strawberry) - the rubbing off of skin by friction.
- Dislocation - "the displacement of one or more bones of a joint from their natural positions.
- Hot spot - a hot or irritated feeling of the skin which occurs just before a blister forms. These can be prevented by using petroleum jelly over friction-prone areas.
- Blister - a raised spot on the skin filled with liquid. These can generally be avoided by applying lubricants such as petroleum jelly to areas of friction, keeping footwear (socks, shoes, boots) in good repair, and wearing the proper size of boot or shoe.
- Shinsplints - a painful injury to the soft tissues and bone in the shin area. These are generally caused by wearing shoes with inflexible soles or inadequate shock absorption, running on the toes or on hard surfaces, and/or having calf muscles with a limited range of motion.
- Sprain - a stretching or tearing of the ligament(s) at a joint.
- Muscle spasm (muscle cramp) - a sudden, involuntary contraction of one or more muscles.
- Contusion - a bruise with bleeding into the muscle tissue.
- Strain - a stretching or tearing of the muscles.
- Bursitis - an inflammation of the bursa (a sack-like structure where tendons pass over bones). This occurs at a joint and produces pain when the joint is moved or touched. Sometimes swelling occurs.
- Tendinitis - an inflammation of a tendon that produces pain when the attached muscle contracts. Swelling may not occur.
- Stress fractures of the feet.
- Tibial stress fractures - overuse injuries which seem like shinsplints except that the pain is in a specific area.
- Knee injuries - caused by running on uneven surfaces or with worn out shoes, overuse, and improper body alignment. Soldiers who have problems with their knees can benefit from doing leg exercises that strengthen the front (quadriceps) and rear (hamstrings) thigh muscles.
- Low back problems - caused by poor running, sitting, or lifting techniques, and by failing to stretch the back and hip-flexor muscles and to strengthen the abdominal muscles.

The most common running injuries occur in the feet, ankles, knees, and legs. Although they are hard to eliminate, much can be done to keep them to a minimum. Preventive measures include proper warm-up and cool-down along with stretching exercises. Failure to allow recovery between hard bouts of running can lead to overtraining and can also be a major cause of injuries. A well-conditioned soldier can run five to six times a week. However, to do this safely, he should do two things: gradually build up to running that frequently and vary the intensity of the running sessions to allow recovery between them.

Many running injuries can be prevented by wearing proper footwear. Soldiers should train in running shoes. These are available in a wide range of prices and styles. They should fit properly and have flexible, multi-layered soles with good arch and heel support. Shoes made with leather and nylon uppers are usually the most comfortable. See Appendix E for more information on running shoes.

Since injuries can also be caused by running on hard surfaces, soldiers should, if possible, avoid running on concrete. Soft, even surfaces are best for injury prevention. Whenever possible, soldiers should run on grass paths, dirt paths, or park trails. However, with adequate footwear and recovery periods, running on roads and other hard surfaces should pose no problem.

Common running injuries include the following:

- Black toenails.
- Ingrown toenails.
- Stress fractures of the feet.
- Ankle sprains and fractures.
- Achilles tendinitis (caused by improper stretching and shoes that do not fit.
- Upper leg and groin injuries (which can usually be prevented by using good technique in stretching and doing strengthening exercises).

Tibial stress fractures, knee injuries, low back problems, shinsplints, and blisters, which were mentioned earlier, are also injuries which commonly occur in runners.

Other Factors

Proper clothing can also help prevent injuries. Clothes used for physical activity should be comfortable and fit loosely. A T-shirt or sleeveless undershirt and gym shorts are best in warm weather. In cold weather, clothing may be layered according to personal preference. For example, soldiers can wear a BDU, sweat suit, jogging suit, or even Army-issued long underwear. In very cold weather, soldiers may need gloves or mittens and ear-protecting caps. Rubberized or plastic suits should never be worn during exercise. They cause excessive sweating which can lead to dehydration and a dangerous increase in body temperature.

Many running injuries can be prevented by wearing proper footwear.

Army Regulation 385-55 (paragraph B- 12, C) prohibits the use of headphones or earphones while walking, jogging, skating, or bicycling on the roads and streets of military installations. However, they may be worn on tracks and running trails.

Road safety equipment is required on administative-type walks, marches, or runs which cross highways, roads, or tank trails or which are conducted on traffic ways. If there is reduced visibility, control personnel must use added caution to ensure the safety of their soldiers.

CHAPTER 14

Army Physical Fitness Test

The APFT is a three-event physical performance test used to assess muscular endurance and cardiorespiratory (CR) fitness .

All soldiers in the Active Army, Army National Guard, and Army Reserve must take the Army Physical Fitness Test (APFT) regardless of their age. The APFT is a three-event physical performance test used to assess muscular endurance and cardiorespiratory (CR) fitness. It is a simple way to measure a soldier's ability to effectively move his body by using his major muscle groups and CR system. Performance on the APFT is strongly linked to the soldier's fitness level and his ability to do fitness-related tasks. An APFT with alternate test events is given to soldiers with permanent profiles and with temporary profiles greater than three months' duration.

While the APFT testing is an important tool in determining the physical readiness of individual soldiers and units, it should not be the sole basis for the unit's physical fitness training. Commanders at every level must ensure that fitness training is designed to develop physical abilities in a balanced way, not just to help soldiers do well on the APFT.

Commanders should use their unit's APFT results to evaluate its physical fitness level. APFT results may indicate a need to modify the fitness programs to attain higher fitness levels. However, mission-essential tasks, not the APFT, should drive physical training.

Performance on the APFT is strongly linked to the soldier's fitness level and his ability to do fitness-related tasks.

Additional physical performance tests and standards which serve as prerequisites for Airborne/Ranger/Special Forces/SCUBA qualification are provided in DA Pam 351-4.

Methods of Evaluation

Commanders are responsible for ensuring that their soldiers are physically fit (AR 350- 15). There are several ways they can assess fitness including the following

- Testing. This is an efficient way to evaluate both the individual's and the unit's physical performance levels.
- Inspection. This evaluates training procedures and indicates the sound ness of the unit's physical fitness program.
- Observation. This is an ongoing way to review training but is not as reliable as testing as an indicator of the unit's level of fitness.
- Medical examination. This detects individual disabilities, health-related problems, and physical problems.

Over-Forty Cardiovascular Screening Program

The Army's over-40 cardiovascular screening program (CVSP) does the following:

- Identifies soldiers with a risk of coronary heart disease.
- Provides guidelines for safe, regular CR exercise.
- Gives advice and help in controlling heart-disease risk factors.
- Uses treadmill testing only for high-risk soldiers who need it.

All soldiers, both active and reserve component, must take the APFT for record regardless of age unless prohibited by a medical profile. For soldiers who reached age 40 on or after 1 January 1989, there is no requirement for clearance in the cardiovascular screening program before taking a record APFT. Soldiers who reached age 40 before 1 January 1989 must be cleared through the cardiovascular screening program before taking a record APFT. Prior to their CVSP evaluation, however, they may still take part in physical training to include diagnostic APFTs unless profiled or contraindications to exercise exist. All soldiers must undergo periodic physical examinations in accordance with AR 40-501 and NGR 40-501. These include screening for cardiovascular risk factors.

Overview

As stated, APFT events assess muscular endurance and CR fitness. The lowest passing APFT standards reflect the minimum acceptable fitness level for all soldiers, regardless of MOS or component. When applied to a command, APFT results show a unit's overall level of physical fitness. However, they are not all-inclusive, overall measures of physical-combat readiness. To assess this, other physical capabilities must be measured. The APFT does, however, give a commander a sound measurement of the general fitness level of his unit.

Service schools, agencies, and units may set performance goals which are above the minimum APFT standards in accordance with their missions (AR 350- 15). Individual soldiers are also encouraged to set for themselves a series of successively higher APFT performance goals. They should always strive to improve themselves physically and never be content with meeting minimum standards. Competition on the APFT among soldiers or units can also be used to motivate them to improve their fitness levels.

Testing is not a substitute for a regular, balanced exercise program. Diagnostic testing is important in monitoring training progress but, when done too often, may decrease motivation and waste training time.

The test period is defined as *the* period of time which elapses from starting to finishing the three events. It must not take more than two hours. Soldiers must do all three events in the same test period.

Test Administration

The APFT must be administered properly and to standard in order to accurately evaluate a soldier's physical fitness and to be fair to all soldiers. (Test results are used for personnel actions.)

Individual soldiers are not authorized to administer the APFT to themselves for the purpose of satisfying a unit's diagnostic or record APFT requirement.

REQUIRED EQUIPMENT

The OIC or NCOIC at the test site must have a copy of FM 21-20 on hand. The supervisor of each event must have the event instructions and standards. Scorers should have a clipboard and an ink pen to record the results on the soldiers' scorecards.

Two stopwatches are needed. They must be able to measure time in both minutes and seconds.

Runners must wear numbers or some other form of identification for the 2-mile run. The numbers may be stenciled or pinned onto pullover vests or sleeveless, mesh pullovers or attached to the runners themselves.

Soldiers should wear clothing that is appropriate for PT such as shorts, T-shirts, socks, and running shoes (not tennis shoes). They should not wear basketball shoes or other types of court shoes. BDUs may be worn but may be a hindrance on some events.

Anything that gives a soldier an unfair advantage is not permitted during the APFT. Wearing devices such as weight belts or elastic bandages may or may not provide an advantage. However, for standardization, such additional equipment is not authorized unless prescribed by medical personnel. The only exception is gloves. They may be worn in cold weather when approved by the local commander.

Each soldier needs a DA Form 705, Army Physical Fitness Test Scorecard. The soldier fills in his name, social security number, grade, age, and sex.

(See Figure 14-1.) The unit will complete the height and weight data.

Scorers record the raw score for each event and initial the results. If a soldier fails an event or finds it difficult to perform, the scorer should write down the reasons and other pertinent information in the comment block. After the entire APFT has been completed, the event scorer will convert raw scores to point scores using the scoring standards on the back of the scorecards. (See Figure 14-1.)

See page 14-8.1 for instructions on completing DA Form 705.

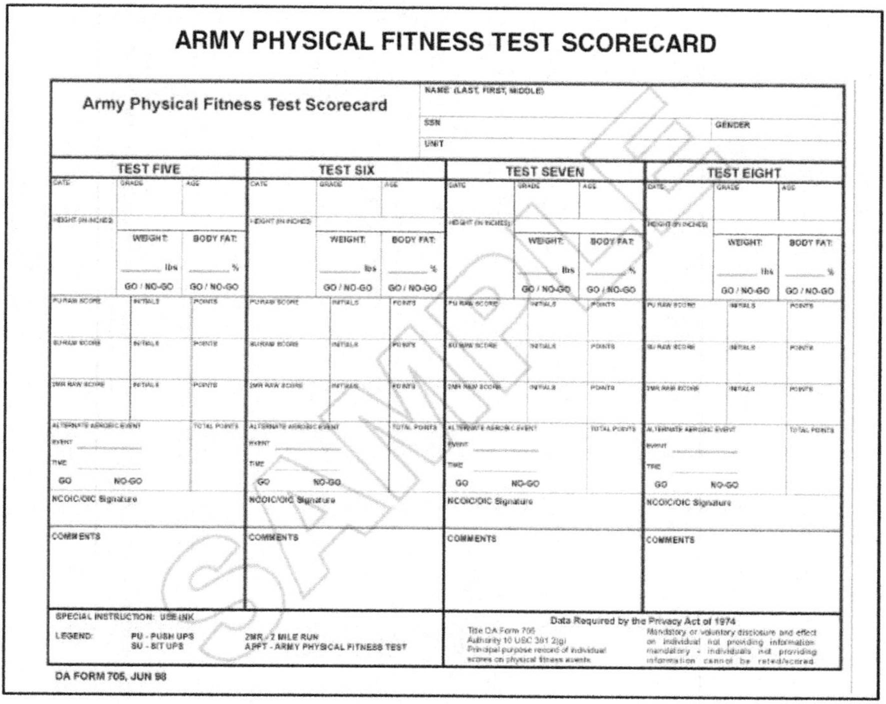

ARMY PHYSICAL FITNESS TEST SCORECARD

Army Physical Fitness Test Scorecard

NAME (LAST, FIRST, MIDDLE)

SSN | GENDER

UNIT

TEST FIVE	TEST SIX	TEST SEVEN	TEST EIGHT
DATE / GRADE / AGE	DATE / GRADE / AGE	DATE / GRADE / AGE	DATE / GRADE / AGE
HEIGHT (IN INCHES)	HEIGHT (IN INCHES)	HEIGHT (IN INCHES)	HEIGHT (IN INCHES)
WEIGHT: ____ lbs GO / NO-GO; BODY FAT: ____ % GO / NO-GO	WEIGHT: ____ lbs GO / NO-GO; BODY FAT: ____ % GO / NO-GO	WEIGHT: ____ lbs GO / NO-GO; BODY FAT: ____ % GO / NO-GO	WEIGHT: ____ lbs GO / NO-GO; BODY FAT: ____ % GO / NO-GO
PU RAW SCORE / INITIALS / POINTS	PU RAW SCORE / INITIALS / POINTS	PU RAW SCORE / INITIALS / POINTS	PU RAW SCORE / INITIALS / POINTS
SU RAW SCORE / INITIALS / POINTS	SU RAW SCORE / INITIALS / POINTS	SU RAW SCORE / INITIALS / POINTS	SU RAW SCORE / INITIALS / POINTS
2MR RAW SCORE / INITIALS / POINTS	2MR RAW SCORE / INITIALS / POINTS	2MR RAW SCORE / INITIALS / POINTS	2MR RAW SCORE / INITIALS / POINTS
ALTERNATE AEROBIC EVENT / TOTAL POINTS	ALTERNATE AEROBIC EVENT / TOTAL POINTS	ALTERNATE AEROBIC EVENT / TOTAL POINTS	ALTERNATE AEROBIC EVENT / TOTAL POINTS
EVENT ____ TIME ____ GO NO-GO	EVENT ____ TIME ____ GO NO-GO	EVENT ____ TIME ____ GO NO-GO	EVENT ____ TIME ____ GO NO-GO
NCOIC/OIC Signature	NCOIC/OIC Signature	NCOIC/OIC Signature	NCOIC/OIC Signature
COMMENTS	COMMENTS	COMMENTS	COMMENTS

SPECIAL INSTRUCTION: USE INK

LEGEND: PU - PUSH UPS; SU - SIT UPS; 2MR - 2 MILE RUN; APFT - ARMY PHYSICAL FITNESS TEST

Data Required by the Privacy Act of 1974

Title DA Form 705
Authority 10 USC 301 2(g)
Principal purpose record of individual scores on physical fitness events
Mandatory or voluntary disclosure and effect on individual not providing information: mandatory - individuals not providing information cannot be rated/scored

DA FORM 705, JUN 98

SAMPLE

***Figure 14-1**

Age Group	17-21		22-26		27-31		32-36		37-41		Age Group	42-46		47-51		52-56		57-61		62+		Age Group
PUSH-UP STANDARDS																						
Repetitions	M	F	M	F	M	F	M	F	M	F	Repetitions	M	F	M	F	M	F	M	F	M	F	Repetitions
77					100						77											77
76					99						76											76
75			100		98		100				75											75
74			99		97		99				74											74
73			98		96		98		100		73											73
72			97		95		97		99		72											72
71	100		96		94		96		98		71											71
70	99		94		93		95		97		70											70
69	97		93		92		94		96		69											69
68	96		92		91		93		95		68											68
67	94		91		89		92		94		67											67
66	93		90		88		91		93		66	100										66
65	92		89		87		90		92		65	99										65
64	90		87		86		89		91		64	98										64
63	89		86		85		88		90		63	97										63
62	88		85		84		87		89		62	96										62
61	86		84		83		86		88		61	94										61
60	85		83		82		85		87		60	93										60
59	83		82		81		84		86		59	92		100								59
58	82		81		80		83		85		58	91		99								58
57	81		79		79		82		84		57	90		98								57
56	79		78		78		81		83		56	89		96		100						56
55	78		77		77		79		82		55	88		95		99						55
54	77		76		76		78		81		54	87		94		98						54
53	75		75		75		77		79		53	86		93		97		100				53
52	74		74		74		76		78		52	84		92		96		99				52
51	72		73		73		76		77		51	83		91		94		98				51
50	71		71		72	100	74		76		50	82		89		93		97		100		50
49	70		70		71	99	73		75		49	81		88		92		96		99		49
48	68		69		69	98	72		74		48	80		87		91		94		98		48
47	67		68		68	96	71		73		47	79		86		90		93		96		47
46	66		67	100	67	95	70		72		46	78		85		89		92		95		46
45	64		66	99	66	94	69	100	71		45	77		84		88		91		94		45
44	63		65	97	65	93	68	99	70		44	76		82		87		90		93		44
43	61		63	96	64	92	67	97	69		43	74		81		86		89		92		43
42	60	100	62	94	63	90	66	96	68		42	73		80		84		87		91		42
41	59	98	61	93	62	89	65	95	67		41	72		79		83		86		89		41
40	57	97	60	92	61	88	64	93	66	100	40	71		78		82		85		88		40
39	56	95	59	90	60	87	63	92	65	99	39	70		76		81		84		87		39
38	54	93	58	89	59	85	62	91	64	97	38	69		75		80		83		86		38
37	53	91	57	88	58	84	61	89	63	96	37	68	100	74		79		82		85		37
36	52	90	55	86	57	83	60	88	62	94	36	67	99	73		78		81		84		36
35	50	88	54	85	56	82	59	87	61	93	35	66	97	72		77		79		82		35
34	49	86	53	83	55	81	58	85	60	91	34	64	95	71	100	76		78		81		34
33	48	84	52	82	54	79	57	84	59	90	33	63	94	69	98	74		77		80		33
32	46	83	51	81	53	78	56	83	58	88	32	62	92	68	97	73		76		79		32
31	45	81	50	79	52	77	55	81	57	87	31	61	90	67	95	72	100	75		78		31
30	43	79	49	78	50	76	54	80	56	85	30	60	89	66	93	71	98	74		76		30
29	42	77	47	77	49	75	53	79	55	84	29	59	87	65	92	70	96	73		75		29
28	41	76	46	75	48	73	52	77	54	82	28	58	86	64	90	69	95	71	100	74		28
27	39	74	45	74	47	72	51	76	53	81	27	57	84	62	88	68	93	70	98	73		27
26	38	72	44	72	46	71	50	75	52	79	26	56	82	61	87	67	91	69	96	72		26
25	37	70	43	71	45	70	49	73	51	78	25	54	81	60	85	66	89	68	94	71	100	25
24	35	69	42	70	44	68	48	72	50	76	24	53	79	59	83	64	87	67	92	69	98	24
23	34	67	41	68	43	67	47	71	49	75	23	52	78	58	82	63	85	66	90	68	96	23
22	32	65	39	67	42	66	46	69	48	73	22	51	76	56	80	62	84	65	88	67	93	22
21	31	63	38	66	41	65	45	68	47	72	21	50	74	55	78	61	82	63	86	66	91	21
20	30	62	37	64	40	64	44	67	46	70	20	49	73	54	77	60	80	62	84	65	89	20
19	28	60	36	63	39	62	43	65	45	69	19	48	71	53	75	59	78	61	82	64	87	19
18	27	58	35	61	38	61	42	64	44	67	18	47	70	52	73	58	76	60	80	62	84	18
17	26	57	34	60	37	60	41	63	43	66	17	46	68	51	72	57	75	59	78	61	82	17
16	24	55	33	59	36	59	39	61	42	64	16	44	66	49	70	56	73	58	76	60	80	16
15	23	53	31	57	35	58	38	60	41	63	15	43	65	48	68	54	71	57	74	59	78	15
14	21	51	30	56	34	56	37	59	39	61	14	42	63	47	67	53	69	55	72	58	76	14
13	20	50	29	54	33	55	36	58	38	60	13	41	62	46	65	52	67	54	70	56	73	13
12	19	48	28	52	32	54	35	56	37	59	12	40	60	45	63	51	65	53	68	55	71	12
11	17	46	27	50	31	52	34	54	36	57	11	39	58	44	62	50	64	52	66	54	69	11
10	16	44	26	49	29	50	33	52	35	56	10	38	57	42	60	49	62	51	64	53	67	10
9	14	43	25	49	28	49	32	50	34	54	9	37	55	41	58	48	60	50	62	52	64	9
8	13	41	23	48	27	48	31	49	33	53	8	36	54	40	57	47	58	49	60	51	62	8
7	12	39	22	46	26	48	30	49	32	51	7	34	52	39	55	46	56	47	58	49	60	7
6	10	37	21	45	25	47	29	48	31	50	6	33	50	38	53	44	55	46	56	48	58	6
5	9	36	20	43	24	45	28	47	30	48	6	32	49	36	52	43	53	45	54	47	56	5
4	8	34	19	42	23	44	27	45	29	47												
3	6	32	18	41	22	43	26	44	28	45												
2	5	30	17	39	21	42	25	43	27	44												
1	3	29	15	38	20	41	24	41	26	42												
Repetitions	M	F	M	F	M	F	M	F	M	F	Repetitions	M	F	M	F	M	F	M	F	M	F	Repetitions
Age Group	17-21		22-26		27-31		32-36		37-41		Age Group	42-46		47-51		52-56		57-61		62+		Age Group

Scoring standards are used to convert raw scores to point scores after test events are completed. Male point scores are indicated by the M at the top and bottom of the shaded column. Female point scores are indicated by the F at the top and bottom of the unshaded column. To convert raw scores to point scores, find the number of repetitions performed in the left-hand column. Next, move right along that row and locate the intersection of the soldier's appropriate age column. Record that number in the Push-Up points block on the front of the scorecard.

***Figure 14-1 (continued)**

SIT-UP STANDARDS

AGE GROUP	17-21	22-26	27-31	32-36	37-41	AGE GROUP	42-46	47-51	52-56	57-61	62+	AGE GROUP
Repetitions	M/F	M/F	M/F	M/F	M/F	Repetitions	M/F	M/F	M/F	M/F	M/F	Repetitions
82			100			82						82
81			99			81						81
80		100	98			80						80
79		99	97			79						79
78	100	97	96			78						78
77	98	96	95			77						77
76	97	95	94	100	100	76						76
75	95	93	92	99	99	75						75
74	94	92	91	98	98	74						74
73	92	91	90	96	97	73						73
72	90	89	89	95	96	72	100					72
71	89	88	88	94	95	71	99					71
70	87	87	87	93	94	70	98					70
69	86	85	86	92	93	69	97					69
68	84	84	85	91	92	68	96					68
67	82	83	84	89	91	67	95					67
66	81	81	83	88	89	66	94	100	100			66
65	79	80	82	87	88	65	93	99	99			65
64	78	79	81	86	87	64	92	98	98	100		64
63	76	77	79	85	86	63	91	97	97	99	100	63
62	74	76	78	84	85	62	90	96	96	98	99	62
61	73	75	77	82	84	61	89	94	95	97	98	61
60	71	73	76	81	83	60	88	93	94	96	97	60
59	70	72	75	80	82	59	87	92	93	95	96	59
58	68	71	74	79	81	58	86	91	92	94	95	58
57	66	69	73	78	80	57	85	90	91	92	94	57
56	65	68	72	76	79	56	84	89	89	91	92	56
55	63	67	71	75	78	55	83	88	88	90	91	55
54	62	65	70	74	77	54	82	87	87	89	90	54
53	60	64	69	73	76	53	81	86	86	88	89	53
52	58	63	68	72	75	52	80	84	85	87	88	52
51	57	61	66	71	74	51	79	83	84	86	87	51
50	55	60	65	69	73	50	78	82	83	85	86	50
49	54	59	64	68	72	49	77	81	82	84	85	49
48	52	57	63	67	71	48	76	80	81	83	84	48
47	50	56	62	66	69	47	75	79	80	82	83	47
46	49	55	61	65	68	46	74	78	79	81	82	46
45	47	53	60	64	67	45	73	77	78	79	81	45
44	46	52	59	62	66	44	72	76	77	78	79	44
43	44	50	58	61	65	43	71	74	76	77	78	43
42	42	49	57	60	64	42	70	73	75	76	77	42
41	41	48	56	59	63	41	69	72	74	75	76	41
40	39	47	55	58	62	40	68	71	73	74	75	40
39	38	45	54	56	61	39	67	70	72	73	74	39
38	36	44	52	55	60	38	66	69	71	72	73	38
37	34	43	51	54	59	37	65	68	69	71	72	37
36	33	41	50	53	58	36	64	67	68	70	71	36
35	31	40	49	52	57	35	63	66	67	69	70	35
34	30	39	48	50	56	34	62	64	66	68	69	34
33	28	37	47	49	55	33	61	63	65	66	68	33
32	26	36	46	48	54	32	60	62	64	65	66	32
31	25	35	45	47	53	31	59	61	63	64	65	31
30	23	33	44	46	52	30	58	60	62	63	64	30
29	22	32	43	45	50	29	57	59	61	62	63	29
28	20	31	42	44	49	28	56	58	60	61	62	28
27	18	29	41	42	48	27	55	57	59	60	61	27
26	17	28	39	41	47	26	54	56	58	59	60	26
25	15	27	38	40	46	25	53	54	57	58	59	25
24	14	25	37	39	45	24	52	53	56	57	58	24
23	12	24	36	38	44	23	51	52	55	56	57	23
22	10	23	35	36	43	22	50	51	54	55	56	22
21	9	21	34	35	42	21	49	50	53	54	55	21
Repetitions	M/F	M/F	M/F	M/F	M/F	Repetitions	M/F	M/F	M/F	M/F	M/F	Repetitions
AGE GROUP	17-21	22-26	27-31	32-36	37-41	AGE GROUP	42-46	47-51	52-56	57-61	62+	AGE GROUP

Scoring standards are used to convert raw scores to point scores after test events are completed. To convert raw scores to point scores, find the number of repetitions performed in the left-hand column. Next, move right along that row and locate the intersection of the soldier's appropriate age column. Record that number in the Sit-Up points block on the front of the scorecard.

*Figure 14-1 (continued)

2-MILE RUN STANDARDS

AGE GROUP	17-21		22-26		27-31		32-36		37-41		AGE GROUP	42-46		47-51		52-56		57-61		62+		AGE GROUP
Time	M	F	M	F	M	F	M	F	M	F	Time	M	F	M	F	M	F	M	F	M	F	Time
12:54											12:54											12:54
13:00	100		100								13:00											13:00
13:06	99		99								13:06											13:06
13:12	97		98								13:12											13:12
13:18	96		97		100		100				13:18											13:18
13:24	94		96		99		99				13:24											13:24
13:30	93		94		98		98				13:30											13:30
13:36	92		93		97		97		100		13:36											13:36
13:42	90		92		96		96		99		13:42											13:42
13:48	89		91		95		96		98		13:48											13:48
13:54	88		90		94		95		97		13:54											13:54
14:00	86		89		92		94		97		14:00											14:00
14:06	85		88		91		93		96		14:06	100										14:06
14:12	83		87		90		92		95		14:12	99										14:12
14:18	82		86		89		91		94		14:18	98										14:18
14:24	81		84		88		90		93		14:24	97		100								14:24
14:30	79		83		87		89		92		14:30	97		99								14:30
14:36	78		82		86		88		91		14:36	96		98								14:36
14:42	77		81		85		87		91		14:42	95		98		100						14:42
14:48	75		80		84		86		90		14:48	94		97		99						14:48
14:54	74		79		83		85		89		14:54	93		96		98						14:54
15:00	72		78		82		85		88		15:00	92		95		98						15:00
15:06	71		77		81		64		87		15:06	91		95		97						15:06
15:12	70		76		79		83		86		15:12	90		94		96						15:12
15:18	68		74		78		82		86		15:18	90		93		95		100				15:18
15:24	67		73		77		81		85		15:24	89		92		95		99				15:24
15:30	66		72		76		80		84		15:30	88		91		94		98				15:30
15:36	64	100	71	100	75		79		83		15:36	87		91		93		97				15:36
15:42	63	99	70	99	74		78		82		15:42	86		90		92		97		100		15:42
15:48	61	98	69	98	73	100	77		81		15:48	85		89		91		96		99		15:48
15:54	60	96	68	97	72	99	76	100	80		15:54	84		88		91		95		98		15:54
16:00	59	95	67	96	71	98	75	99	80		16:00	83		87		90		94		97		16:00
16:06	57	94	66	95	70	97	75	99	79		16:06	83		87		89		93		96		16:06
16:12	56	93	64	94	69	97	74	98	78		16:12	82		86		88		92		95		16:12
16:18	54	92	63	93	68	96	73	97	77		16:18	81		85		87		91		94		16:18
16:24	53	90	62	92	66	95	72	97	76		16:24	80		84		87		90		93		16:24
16:30	52	89	61	91	65	94	71	96	75		16:30	79		84		86		90		93		16:30
16:36	50	88	60	90	64	93	70	95	74		16:36	78		83		85		89		92		16:36
16:42	49	87	59	89	63	92	69	94	74		16:42	77		82		84		88		91		16:42
16:48	48	85	58	88	62	91	68	94	[illegible]		16:48	77		81		84		87		90		16:48
16:54	46	84	57	87	61	91	67	93	72		16:54	76		80		83		86		89		16:54
17:00	45	83	56	86	60	90	66	92	71	100	17:00	75		80		82		85		88		17:00
17:06	43	82	54	85	59	89	65	92	70	99	17:06	74		79		81		84		87		17:06
17:12	42	81	53	84	58	88	65	91	69	99	17:12	73		78		80		83		86		17:12
17:18	41	79	52	83	57	87	64	90	69	98	17:18	72		77		80		83		85		17:18
17:24	39	78	51	82	56	86	63	90	68	97	17:24	71	100	76		79		82		84		17:24
17:30	38	77	50	81	55	86	62	89	67	96	17:30	70	99	76		78		81		83		17:30
17:36	37	76	49	80	54	85	61	88	66	96	17:36	70	99	75	100	77		80		82		17:36
17:42	35	75	48	79	52	84	60	88	65	95	17:42	69	98	74	99	76		79		81		17:42
17:48	34	73	47	78	51	83	59	87	64	94	17:48	68	97	73	99	76		78		80		17:48
17:54	32	72	46	77	50	82	58	86	63	94	17:54	67	97	73	98	75		77		80		17:54
18:00	31	71	44	76	49	81	57	86	63	93	18:00	66	96	72	97	74		77		79		18:00
18:06	30	70	43	75	48	80	56	85	62	92	18:06	65	96	71	97	73		76		78		18:06
18:12	28	68	42	74	47	80	55	84	61	92	18:12	64	95	70	96	73		75		77		18:12
18:18	27	67	41	73	46	79	55	83	60	91	18:18	63	94	69	96	72		74		76		18:18
18:24	26	66	40	72	45	78	54	83	59	90	18:24	63	94	69	95	71		73		75		18:24
18:30	24	65	39	71	44	77	53	82	58	89	18:30	62	93	68	94	70		72		74		18:30
18:36	23	64	38	70	43	76	52	81	57	89	18:36	61	92	67	94	69		71		73		18:36
18:42	21	62	37	69	42	75	51	81	57	88	18:42	60	92	66	93	69		70		72		18:42
18:48	20	61	36	68	41	74	50	80	56	87	18:48	59	91	65	92	68		70		71		18:48
18:54	19	60	34	67	39	74	49	79	55	87	18:54	58	90	65	92	67		69		70		18:54
19:00	17	59	33	66	38	73	48	79	54	86	19:00	57	90	64	91	66	100	68		69		19:00
19:06	16	58	32	65	37	72	47	78	53	85	19:06	57	89	63	91	65	99	67		68		19:06
19:12	14	56	31	64	36	71	46	77	52	85	19:12	56	89	62	90	65	99	66		67		19:12
19:18	13	55	30	63	35	70	45	77	51	84	19:18	55	88	62	89	64	98	65		67		19:18
19:24	12	54	29	62	34	69	45	76	51	83	19:24	54	87	61	89	63	97	64		66		19:24
19:30	10	53	28	61	33	69	44	75	50	82	19:30	53	87	60	88	62	96	63		65		19:30
19:36	9	52	27	60	32	68	43	74	49	82	19:36	52	86	59	87	62	96	63		64		19:36
19:42	8	50	26	59	31	67	42	74	48	81	19:42	51	85	58	87	61	95	62	100	63		19:42
19:48	6	49	24	58	30	66	41	73	47	80	19:48	50	85	58	86	60	94	61	99	62		19:48
19:54	5	48	23	57	29	65	40	72	46	80	19:54	50	84	57	86	59	93	60	98	61		19:54
20:00	3	47	22	56	28	64	39	72	46	79	20:00	49	83	56	85	58	93	59	98	60	100	20:00
20:06	[illegible]	45	21	55	[illegible]	63	38	71	[illegible]	78	20:06	48	[illegible]	55	84	58	[illegible]	58	97	59	[illegible]	20:06

*Figure 14-1 (continued)

[illegible]	3	[illegible]		[illegible]	28	[illegible]		[illegible]	46	[illegible]		[illegible]	83	[illegible]		[illegible]	93	[illegible]				[illegible]
20:06	2	45	21	55	26	63	38	71	45	78	20:06	48	83	55	84	58	92	58	97	59	99	20:06
20:12	1	44	20	54	25	63	37	70	44	78	20:12	47	82	55	84	57	91	57	96	58	98	20:12
20:18	0	43	19	53	24	62	36	70	43	77	20:18	46	82	54	83	56	90	57	95	57	98	20:18
20:24		42	18	52	23	61	35	69	42	76	20:24	45	81	53	82	55	90	56	95	56	97	20:24
20:30		41	17	51	22	60	35	68	41	75	20:30	44	80	52	82	55	89	55	94	55	96	20:30
20:36		39	16	50	21	59	34	68	40	75	20:36	43	80	51	81	54	88	54	93	54	95	20:36
20:42		38	14	49	20	58	33	67	40	74	20:42	43	79	51	81	53	87	53	92	53	94	20:42
20:48		37	13	48	19	57	32	66	39	73	20:48	42	78	50	80	52	87	52	91	53	94	20:48
20:54		36	12	47	18	57	31	66	38	73	20:54	41	78	49	79	51	86	51	91	52	93	20:54
21:00		35	11	46	17	56	30	65	37	72	21:00	40	77	48	79	51	85	50	90	51	92	21:00
21:06		33	10	45	16	55	29	64	36	71	21:06	39	77	47	78	50	84	50	89	50	91	21:06
21:12		32	9	44	15	54	28	63	35	71	21:12	38	76	47	77	49	84	49	88	49	90	21:12
21:18		31	8	43	14	53	27	63	34	70	21:18	37	75	46	77	48	83	48	87	48	90	21:18
21:24		30	7	42	12	52	26	62	34	69	21:24	37	75	45	76	47	82	47	87	47	89	21:24
21:30		28	6	41	11	51	25	61	33	68	21:30	36	74	44	76	47	81	46	86	46	88	21:30
21:36		27	4	40	10	51	25	61	32	68	21:36	35	73	44	75	46	81	45	85	45	87	21:36
21:42		26	3	39	9	50	24	60	31	67	21:42	34	73	43	74	45	80	44	84	44	86	21:42
21:48		25	2	38	8	49	23	59	30	66	21:48	33	72	42	74	44	79	43	84	43	86	21:48
21:54		24	1	37	7	48	22	59	29	66	21:54	32	71	41	73	44	79	43	83	42	85	21:54
22:00		22	0	36	6	47	21	58	29	65	22:00	31	71	40	72	45	78	42	82	41	84	22:00
22:06		21		35	5	46	20	57	28	64	22:06	30	70	40	72	42	77	41	81	40	83	22:06
22:12		20		34	4	46	19	57	27	64	22:12	30	70	39	71	41	76	40	80	40	82	22:12
22:18		19		33	3	45	18	56	26	63	22:18	29	69	38	71	40	76	39	80	39	82	22:18
22:24		18		32	2	44	17	55	25	62	22:24	28	68	37	70	40	75	38	79	38	81	22:24
22:30		16		31	1	43	16	54	24	61	22:30	27	68	36	69	39	74	37	78	37	80	22:30
22:36		15		30	0	42	15	54	23	61	22:36	26	67	36	69	38	73	37	77	36	79	22:36
22:42		14		29		41	15	53	23	60	22:42	25	66	35	68	37	73	[illegible]	76	35	78	22:42
22:48		13		28		40	14	52	22	59	22:48	24	66	34	67	36	72	35	76	34	78	22:48
22:54		12		27		40	13	52	21	59	22:54	23	65	33	67	36	71	34	75	33	77	22:54
23:00		10		26		39	12	51	20	58	23:00	23	64	33	66	35	70	33	74	32	76	23:00
23:06		9		25		38	11	50	19	57	23:06	22	64	32	66	34	70	32	73	31	75	23:06
23:12		8		24		37	10	49	18	56	23:12	21	63	31	65	33	69	31	73	30	74	23:12
23:18		7		23		36	9	49	17	56	23:18	20	63	30	64	33	68	30	72	29	74	23:18
23:24		5		22		35	8	48	17	55	23:24	19	62	29	64	32	67	30	71	28	73	23:24
23:30		4		21		34	7	48	16	54	23:30	18	61	29	63	31	67	29	70	27	72	23:30
23:36		3		20		34	6	47	15	54	23:36	17	61	28	62	30	66	28	69	27	71	23:36
23:42		2		19		33	5	46	14	53	23:42	17	60	27	62	29	65	27	69	26	70	23:42
23:48		1		18		32	5	46	[illegible]	52	23:48	16	59	26	61	29	64	26	68	25	70	23:48
23:54		0		17		31	4	45	12	[illegible]	23:54	15	59	25	61	28	64	25	67	24	69	23:54
24:00				16		30	3	44	[illegible]	51	24:00	14	58	25	60	27	63	24	66	23	68	24:00
24:06				15		29	2	43	11	50	24:06	13	57	24	59	26	62	23	65	22	67	24:06
24:12				14		29	1	43	10	49	24:12	12	57	23	59	25	61	23	65	21	66	24:12
24:18				13		28	0	42	9	49	24:18	11	56	22	58	25	61	22	64	20	66	24:18
24:24				12		27		41	8	48	24:24	10	56	22	57	24	60	21	63	19	65	24:24
24:30				11		26		41	7	47	24:30	10	55	21	57	23	59	20	62	18	64	24:30
24:36				10		25		40	6	47	24:36	9	54	20	56	22	59	19	62	17	63	24:36
24:42				9		24		39	6	46	24:42	8	54	19	56	22	58	18	61	16	62	24:42
24:48				8		23		39	5	45	24:48	7	53	18	55	21	57	17	60	15	62	24:48
24:54				7		23		38	4	45	24:54	6	52	18	54	20	56	17	59	14	61	24:54
25:00				6		22		37	3	44	25:00	5	52	17	54	19	56	16	58	13	60	25:00
25:06				5		21		37	2	43	25:06	4	51	16	53	18	55	15	58	13	59	25:06
25:12				4		20		36	1	42	25:12	3	50	15	52	18	54	14	57	12	58	25:12
25:18				3		19		35	0	42	25:18	3	50	15	52	17	53	13	56	11	58	25:18
25:24				2		18		34		41	25:24	2	49	14	51	16	53	12	55	10	57	25:24
25:30				1		17		34		40	25:30	1	49	13	51	15	52	11	55	9	56	25:30
25:36				0		17		33		40	25:36	0	48	12	50	15	51	10	54	8	55	25:36
25:42						16		32		39	25:42		47	11	49	14	50	10	53	7	54	25:42
25:48						15		32		38	25:48		47	11	49	13	50	9	52	6	54	25:48
25:54						14		31		38	25:54		46	10	48	12	49	8	51	5	53	25:54
26:00						13		30		37	26:00		45	9	47	11	48	7	51	4	52	26:00
26:06						12		30		36	26:06		45	8	47	11	47	6	50	3	51	26:06
26:12						11		29		35	26:12		44	7	46	10	47	5	49	2	50	26:12
26:18						11		28		35	26:18		43	7	46	9	46	4	48	1	50	26:18
26:24						10		28		34	26:24		43	6	45	8	45	3	47	0	49	26:24
26:30						9		27		33	26:30		42	5	44	7	44	3	47	0	48	26:30
Time	M	F	M	F	M	F	M	F	M	F	Time	M	F	M	F	M	F	M	F	M	F	Time
AGE GROUP	17-21		22-26		27-31		32-36		37-41		AGE GROUP	42-46		47-51		52-56		57-61		62+		AGE GROUP

Scoring standards are used to convert raw scores to point scores after test events are completed. Male point scores are indicated by the M at the top and bottom of the shaded column. Female point scores are indicated by the F at the top and bottom of the unshaded column. To convert raw scores to point scores, find the number of repetitions performed in the left-hand column. Next, move right along that row and locate the intersection of the soldier's appropriate age column. **In all cases, when a time falls between two point values, the lower point value is used.** Record that number in the 2MR points block on the front of the scorecard.

***Figure 14-1 (continued)**

SUPERVISION

The APFT must be properly supervised to ensure that its objectives are met. Proper supervision ensures uniformity in the following:

- Scoring the test.
- Training of supervisors and scorers.
- Preparing the test and controlling performance factors.

The goal of the APFT is to get an accurate evaluation of the soldiers' fitness levels. Preparations for administering an accurate APFT include the following:

- Selecting and training supervisors and scorers.
- Briefing and orienting administrators and participants.
- Securing a location for the events.

Commanders must strictly control those factors which influence test performance. They must ensure that events, scoring, clothing, and equipment are uniform. Commanders should plan testing which permits each soldier to perform to his maximal level. They should also ensure the following:

- Soldiers are not tested when fatigued or ill.
- Soldiers do not have tiring duties just before taking the APFT.
- Weather and environmental conditions do not inhibit performance.
- Safety is the first consideration.

Duties of Test Personnel

Testers must be totally familiar with the instructions for each event and trained to administer the tests. Correctly supervising testees and laying out the test area are essential duties. The group administering the test must include the following:

- OIC or NCOIC.
- Event supervisor, scorers, and a demonstrator for each event.
- Support personnel (safety, control, and medical as appropriate). There should be no less than one scorer for each 15 soldiers tested. Twelve to 15 scorers are required when a company-sized unit is tested.

OIC OR NCOIC

The OIC or NCOIC does the following:

- Administers the APFT.
- Procures all necessary equipment and supplies.
- Arranges and lays out the test area.
- Trains the event supervisors, scorers, and demonstrators. (Training video tape No. 21-191 should be used for training those who administer the APFT.)
- Ensures the test is properly administered and the events are explained, demonstrated, and scored according to the test standards in this chapter.
- Reports the results after the test.

EVENT SUPERVISORS

Event supervisors do the following:

- Administer the test events.
- Ensure that necessary equipment is on hand.
- Read the test instructions, and have the events demonstrated.
- Supervise the scoring of events, and ensure that they are done correctly.
- Rule on questions and scoring discrepancies for their event.

SCORERS

Scorers do the following:

- Supervise the performance of testees.
- Enforce the test standards in this chapter.
- Count the number of correctly performed repetitions aloud.
- Record the correct, raw score on each soldier's scorecard, and initial the scorecard block.
- Perform other duties assigned by the OIC or NCOIC.

Scorers must be thoroughly trained to maintain uniform scoring standards. They do not participate in the test.

The goal of the APFT is to get an accurate evaluation of the soldier's fitness levels.

***Instructions for Completing DA Form 705, Army Physical Fitness Scorecard, June 1998.**

NAME Print soldier's last name, first name and middle initial in NAME block.

SSN Print soldier's social security number in SSN block.

GENDER Print **M** for male or **F** for female in GENDER block.

UNIT Print soldier's unit designation in UNIT block.

DATE Print date the APFT is administered in DATE block.

GRADE Print soldier's grade in GRADE block.

AGE Print soldier's age on the date the APFT is administered in AGE block.

HEIGHT Print soldier's height in HEIGHT block. Height will be rounded to the nearest inch. If the height fraction is less than 1/2 inch, round down to the nearest whole number in inches. If the height fraction is greater than 1/2 inch, round up to the next highest whole number in inches.

WEIGHT Print soldier's weight in WEIGHT block. Weight will be recorded to the nearest pound. If the weight fraction is less than 1/2 pound, round down to the nearest pound. If the weight fraction is 1/2 pound or greater, round up to the nearest pound. Circle **GO** if soldier meets screening table weight IAW AR 600-9. Circle **NO-GO** if soldier exceeds screening table weight IAW AR 600-9.

BODY FAT If soldier exceeds screening table weight, print the soldier's body fat in the BODY FAT block. Percent body fat is recorded from DA Form 5500-R, Body Fat Content Worksheet, Dec 85, for male soldiers and DA Form 5501-R, Body Fat Content Worksheet, Dec 85, for female soldiers. Circle **GO** if soldier meets percent body fat for their age and gender IAW AR 600-9. Circle **NO-GO** if soldier exceeds percent body fat for their age and gender IAW AR 600-9. If soldier does not exceed screening table weight or does not appear to have excessive body fat IAW AR 600-9, print N/A (not applicable) in the BODY FAT block.

PU RAW SCORE The event scorer records the number of correctly performed repetitions of the push-up in the PU RAW SCORE block and prints his or her initials in the INITIALS block.

SU RAW SCORE The event scorer records the number of correctly performed repetitions of the sit-up in the SU RAW SCORE block and prints his or her initials in the INITIALS block.

2MR RAW SCORE The event scorer records the two-mile run time in the 2MR RAW SCORE block. The time is recorded in minutes and seconds. The event scorer then determines the point value for the two-mile run using the scoring standards on the reverse side of the scorecard. The point value is recorded in the 2MR POINTS block and the event scorer prints his or her initials in the INITIALS block. In all cases when a point value falls between two point values, the lower point value is used and recorded. The two-mile run event scorer also determines the point value for push-ups and sit-ups using the scoring standards on the reverse side of the scorecard. The point values are recorded in the appropriate push-up and sit-up POINTS block and the event scorer prints his or her initials in the INITIALS block. The two-mile run event scorer totals the points from the three events and records the total APFT score in the TOTAL POINTS block.

ALTERNATE AEROBIC EVENT The event scorer prints the alternate aerobic event administered (800-yard swim, 6.2-mile-stationary bicycle ergometer, 6.2-mile-bicycle test or 2.5-mile walk) in the ALTERNATE AEROBIC EVENT block. The time the soldier completes the alternate aerobic event is recorded in minutes and seconds in the ALTERNATE AEROBIC EVENT block. The standards for the alternate aerobic event tests are listed in FM 21-20, Chapter 14, Figure 14-9. Scoring for all alternate aerobic events is on a **GO** or **NO-GO** basis. No point values are awarded. Circle **GO** if the soldier completes the alternate aerobic event within the required time or less. Circle **NO-GO** if the soldier fails to complete the alternate aerobic event within the required time. The alternate aerobic event scorer also determines the point value for push-ups and or sit-ups using the scoring standards on the reverse side of the scorecard. The point values are recorded in the appropriate push-up and or situp POINTS block and the event scorer prints his or her initials in the 2MR INITIALS block. The alternate aerobic event scorer totals the points from the push-up and or sit-up events and records the total APFT score in the TOTAL POINTS block.

NCOIC/OIC Signature The NCOIC/OIC checks all test scores for accuracy and signs their name in the NCOIC/OIC Signature block.

COMMENTS The event supervisor, event scorer, NCOIC, or OIC may record comments appropriate to the APFT in the COMMENTS block. Appropriate comments may include: weather conditions, injury during APFT and or appeals.

(THIS PAGE INTENTIONALLY LEFT BLANK.)

SUPPORT PERSONNEL

Safety and control people should be at the test site, depending on local policy and conditions. Medical personnel may also be there. However, they do not have to be on site to have the APFT conducted. At a minimum, the OIC or NCOIC should have a plan, known to all test personnel, for getting medical help if needed.

Test Site

The test site should be fairly flat and free of debris. It should have the following:

- An area for stretching and warming up.
- A soft, flat, dry area for performing push-ups and sit-ups.
- A flat, 2-mile running course with a solid surface and no more than a three-percent grade. (Commanders must use good judgement; no one is expected to survey terrain.)
- No significant hazards, (for example, traffic, slippery road surfaces, heavy pollution).

When necessary or expedient, a quarter-mile running track can "be used. It can be marked with a series of stakes along the inside edge. When the track is laid out, a horizontal midline 279 feet, 9 3/4 inches long must be marked in the center of a clear area. A 120-foot circle is marked at both ends of this line. The track is formed when the outermost points of the two circles are connected with tangent lines. (See Figure 14-2.)

RUNNING TRACK

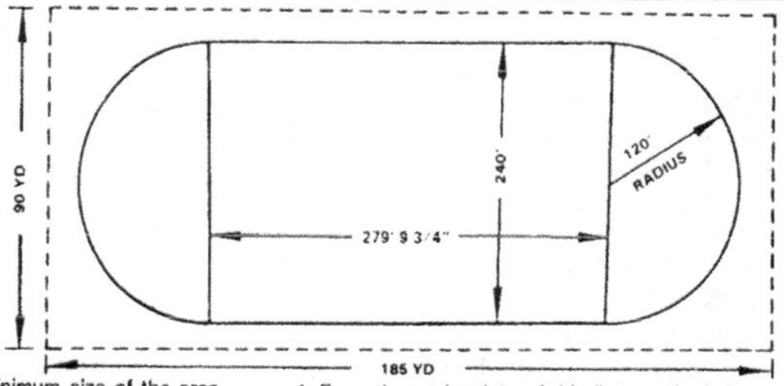

Notes on Layout of Track

1. Dotted lines indicate the minimum size of the area.
2. The track is one-quarter mile at its inside edge.
3. To lay out a track, find a horizontal midline in the center of the area. This line is 279 feet, 9 3/4 inches long.
4. From the end points of this line, mark circles with a radius of 120 feet.
5. To form the track, connect the outermost points of the two circles with the tangent lines.

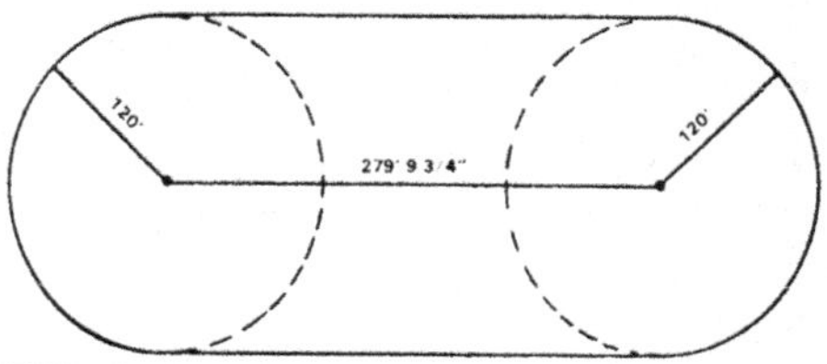

Figure 14-2

A 400-meter track may be used in place of the standard quarter-mile (440-yard) track for the 2-mile run, However, one lap run on a 400-meter track is 92 inches shorter than one lap on a 440-yard track. Eight laps on a 400-meter track is 736 inches shorter than eight laps (2 miles) on a 440-yard track. Therefore, soldiers who run the 2-mile event on a 400-meter track must run eight laps plus an additional 61 feet, 4 inches.

Test Procedures

On test day, soldiers are assembled in a common area and briefed by the test OIC or NCOIC about the purpose and organization of the test. The OIC or NCOIC then explains the scorecard, scoring standards, and sequence of events.

The instructions printed here in large type must be read to the soldiers: "YOU ARE ABOUT TO TAKE THE ARMY PHYSICAL FITNESS TEST, A TEST THAT WILL MEASURE YOUR MUSCULAR ENDURANCE AND CARDIORESPIRATORY FITNESS. THE RESULTS OF THIS TEST WILL GIVE YOU AND YOUR COMMANDERS AN INDICATION OF YOUR STATE OF FITNESS AND WILL ACT AS A GUIDE IN DETERMINING YOUR PHYSICAL TRAINING NEEDS. LISTEN CLOSELY TO THE TEST INSTRUCTIONS, AND DO THE BEST YOU CAN ON EACH OF THE EVENTS."

If scorecards have not already been issued, they are handed out at this time. The OIC or NCOIC then says the following: "IN THE APPROPRIATE SPACES, PRINT IN INK THE PERSONAL INFORMATION REQUIRED ON THE SCORECARD." (If scorecards have been issued to the soldiers and filled out before they arrive at the test site, this remark is omitted.)

The OIC or NCOIC pauses briefly to give the soldiers time to check the information. He then says the following: "YOU ARE TO CARRY THIS CARD WITH YOU TO EACH EVENT. BEFORE YOU BEGIN, HAND THE CARD TO THE SCORER. AFTER YOU COMPLETE THE EVENT, THE SCORER WILL RECORD YOUR RAW SCORE, INITIAL THE CARD, AND RETURN IT TO YOU." (At this point, the scoring tables are explained so everyone understands how raw scores are converted to point scores.) Next, the OIC or NCOIC says the following "EACH OF YOU WILL BE ASSIGNED TO A GROUP. STAY WITH YOUR TEST GROUP FOR THE ENTIRE TEST, WHAT ARE YOUR QUESTIONS ABOUT THE TEST AT THIS POINT?"

Groups are organized as required and given final instructions including what to do after the final event. The test is then given.

RETAKING OF EVENTS

Soldiers who start an event incorrectly must be stopped by the scorer before they complete 10 repetitions and told what their errors are. They are then sent to the end of the line to await their turn to retake the event.

A soldier who has problems such as muscle cramps while performing an event may rest if he does not assume an illegal position in the process. If he continues, he receives credit for all correctly done repetitions within the two-minute period. If he does not continue, he gets credit for the number of correct repetitions he has performed up to that time. If he has not done 10 correct repetitions, he is sent to the end of the line to retake that event. He may not retake the event if he has exceeded 10 repetitions. Soldiers who are unable to perform 10 correct repetitions because of low fitness levels may not retake an event.

TEST FAILURES

Soldiers who stop to rest in an authorized rest position continue to receive credit for correct repetitions performed after their rest. Soldiers who rest in an unauthorized rest position will have their performance in that event immediately terminated.

The records of soldiers who fail a record APFT for the first time and those who fail to take the APFT within the required period (AR 350-15, paragraph 11) must be flagged IAW AR 600-8-2 (Reference B).

RETESTING

Soldiers who fail any or all of the events must retake the entire APFT. In case of test failure, commanders may allow soldiers to retake the test as soon as the soldiers and commanders feel they are ready. Soldiers without a medical profile will be retested not-later-than three months following the initial APFT failure in accordance with AR 350-15, paragraph 11.

Test Sequence

The test sequence is the push-up, sit-up, and 2-mile run (or alternate, aerobic event). The order of events cannot be changed. There are no exceptions to this sequence.

Soldiers should be allowed no less than 10 minutes, but ideally no more than 20 minutes, to recover between each event. The OIC or NCOIC determines the time to be allotted between events, as it will depend on the total number of soldiers who are participating in the APFT. If many soldiers are to be tested, staggered starting times should be planned to allow the proper intervals between events. Under no circumstances is the APFT valid if a soldier cannot begin and end all three events in two hours or less.

The following paragraphs describe the equipment, facilities, personnel, instructions, administration, timing techniques, and scorers' duties for the pushup, sit-up, and 2-mile-run events.

PUSH-UPS

Push-ups measure the endurance of the chest, shoulder, and triceps muscles. (See Figure 14-3.)

Equipment

One stopwatch is needed along with one clipboard and pen for each scorer. The event supervisor must have the following the instructions in this chapter on how to conduct the event and one copy of the push-up scoring standards (DA Form 705).

Facilities

There must be at least one test station for every 15 soldiers to be tested. Each station is 6 feet wide and 15 feet deep.

Personnel

One event supervisor must beat the test site and one scorer at each station. The event supervisor may not be the event scorer.

Instructions

The event supervisor must read the following: "THE PUSH-UP EVENT MEASURES THE ENDURANCE OF THE CHEST, SHOULDER, AND TRICEPS MUSCLES. ON THE COMMAND 'GET SET,' ASSUME THE FRONT-LEANING REST POSITION BY PLACING YOUR HANDS WHERE THEY ARE COMFORTABLE FOR YOU. YOUR FEET MAY BE TOGETHER OR UP TO 12 INCHES APART. WHEN VIEWED FROM THE SIDE, YOUR BODY SHOULD FORM A GENERALLY STRAIGHT LINE FROM YOUR SHOULDERS TO YOUR ANKLES.

PUSH-UPS

Figure 14-3

ON THE COMMAND 'GO,' BEGIN THE PUSH-UP BY BENDING YOUR ELBOWS AND LOWERING YOUR ENTIRE BODY AS A SINGLE UNIT UNTIL YOUR UPPER ARMS ARE AT LEAST PARALLEL TO THE GROUND. THEN, RETURN TO THE STARTING POSITION BY RAISING YOUR ENTIRE BODY UNTIL YOUR ARMS ARE FULLY EXTENDED. YOUR BODY MUST REMAIN RIGID IN A GENERALLY STRAIGHT LINE AND MOVE AS A UNIT WHILE PERFORMING EACH REPETITION. AT THE END OF EACH REPETITION, THE SCORER WILL STATE THE NUMBER OF REPETITIONS YOU HAVE COMPLETED CORRECTLY. IF YOU FAIL TO KEEP YOUR BODY GENERALLY STRAIGHT, TO LOWER YOUR WHOLE BODY UNTIL YOUR UPPER ARMS ARE AT LEAST PARALLEL TO THE GROUND, OR TO EXTEND YOUR ARMS COMPLETELY, THAT REPETITION WILL NOT COUNT, AND THE SCORER WILL REPEAT THE NUMBER OF THE LAST CORRECTLY PERFORMED REPETITION. IF YOU FAIL TO PERFORM THE FIRST TEN PUSH-UPS CORRECTLY, THE SCORER WILL TELL YOU TO GO TO YOUR KNEES AND WILL EXPLAIN TO YOU WHAT YOUR MISTAKES ARE. YOU WILL THEN BE SENT TO THE END OF THE LINE TO BE RETESTED. AFTER THE FIRST 10 PUSH-UPS HAVE BEEN PERFORMED AND COUNTED, HOWEVER, NO RESTARTS ARE ALLOWED. THE TEST WILL CONTINUE, AND ANY INCORRECTLY PERFORMED PUSH-UPS WILL NOT BE COUNTED. AN ALTERED, FRONT-LEANING REST POSITION IS THE ONLY AUTHORIZED REST POSITION. THAT IS, YOU MAY

SAG IN THE MIDDLE OR FLEX YOUR BACK. WHEN FLEXING YOUR BACK, YOU MAY BEND YOUR KNEES, BUT NOT TO SUCH AN EXTENT THAT YOU ARE SUPPORTING MOST OF YOUR BODY WEIGHT WITH YOUR LEGS. IF THIS OCCURS, YOUR PERFORMANCE WILL BE TERMINATED. YOU MUST RETURN TO, AND PAUSE IN, THE CORRECT STARTING POSITION BEFORE CONTINUING. IF YOU REST ON THE GROUND OR RAISE EITHER HAND OR FOOT FROM THE GROUND, YOUR PERFORMANCE WILL BE TERMINATED. YOU MAY REPOSITION YOUR HANDS AND/OR FEET DURING THE EVENT AS LONG AS THEY REMAIN IN CONTACT WITH THE GROUND AT ALL TIMES. CORRECT PERFORMANCE IS IMPORTANT. YOU WILL HAVE TWO MINUTES IN WHICH TO DO AS MANY PUSH-UPS AS YOU CAN. WATCH THIS DEMONSTRATION." (The exercise is then demonstrated. See Figure 14-4 for a list of points that need to be made during the demonstration.) "WHAT ARE YOUR QUESTIONS?"

Administration

After reading the instructions, the supervisor answers questions. Then he moves the groups to their testing stations. The event supervisor cannot be ready to begin. Successive groups do the event until all soldiers have completed it.

Timing Techniques

The event supervisor is the timer. He calls out the time remaining every 30 seconds and every second for the last 10 seconds of the two minutes. He ends the event after two minutes by the command "Halt!"

Scorers' Duties

Scorers must allow for differences in the body shape and structure of each soldier. The scorer uses each soldier's starting position as a guide throughout the event to evaluate each repetition. The scorer should talk to the soldier before the event begins and have him do a few repetitions as a warm-up and reference to ensure he is doing the exercise correctly.

ADDITIONAL POINTS TO DEMONSTRATE FOR THE PUSH-UP EVENT

The following points must be clarified during the demonstration:

- The soldier's chest may touch the ground (mat or floor) during the push-up as long as the contact does not provide him an advantage. He cannot use the ground to bounce off of or momentarily rest on. However, penalizing a soldier for touching the ground with the chest is unfair. Some soldiers have a large chest or abdomen or are otherwise developed in a way which makes touching the ground unavoidable when they are in the correct down position. Do not count those repetitions in which bouncing off the ground has given the soldier an unfair advantage. Do not count those repetitions in which the long bone of the upper arm does not reach a position parallel to the ground.
- Soldiers may reposition their hands during the push-up event as long as the hands remain in contact with the at all times. The hands can be repositioned either forward, inward, outward, or backward. If a soldier repositions his hands too far backward, the legal front-leaning rest position may be violated.
- If a mat is used, the entire body must be on the mat.
- In the rest position, a soldier may sag in the middle or flex his back in the altered front-leaning rest position; however, he may not readjust his hands backward and/or bend his knees to such a point that when he bends at the waist and/or knees, he supports most of his body weight with his legs. If this occurs, the soldier's performance in the event will be terminated.
- The feet may not be braced during the push-up event. Test administrators must ensure that a non-slip surface is available.
- Soldiers may do the push-up event on their fists. This may be necessary due to a prior injury. There is no unfair advantage to be gained by doing so.
- Soldiers may not cross their feet while doing the push-up event. This ensures as much standardization as possible and avoids violation of the proper front-leaning rest position, which is the only authorized starting position for this event.
- Soldiers may not take any part of the APFT in bare feet.
- Soldiers should not wear glasses while performing the push-up event.

Figure 14-4

The scorer may either sit or kneel about three feet from the testee's shoulder at a 45-degree angle in front of it. The scorer's head should be about even with the testee's shoulder when the latter is in the front-leaning rest position. Each scorer determines for himself if he will sit or kneel when scoring. He may not lie down or stand while scoring. He counts out loud the number of correct repetitions completed and repeats the number of the last correct push-up if an incorrect one is done. Scorers tell the testees what they do wrong as it occurs during the event. A critique of the performance is done following the test.

When the soldier completes the event, the scorer records the number of correctly performed repetitions, initials the scorecard, and returns it to the soldier.

SIT-UPS

This event measures the endurance of the abdominal and hip-flexor muscles. (See Figure 14-5.)

Equipment

One stopwatch is needed along with one clipboard and pen for each scorer. The event supervisor must have the following: the instructions in this chapter on how to conduct the event and one copy of the sit-up scoring standards (DA Form 705).

Facilities

Each station is 6 feet wide and 15 feet deep. Ensure that no more than 15 soldiers are tested at a station.

Personnel

One event supervisor must be at the test site and one scorer at each station. The event supervisor may not be the event scorer.

Instructions

The event supervisor must read the following: "THE SIT-UP EVENT MEASURES THE ENDURANCE OF THE ABDOMINAL AND HIP-FLEXOR MUSCLES. ON THE

SIT-UPS

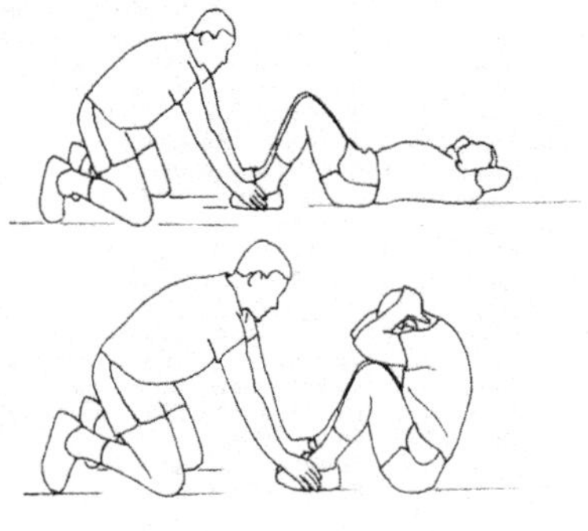

Figure 14-5

COMMAND "GET SET", ASSUME THE STARTING POSITION BY LYING ON YOUR BACK WITH YOUR KNEES BENT AT A 90-DEGREE ANGLE. YOUR FEET MAY BE TOGETHER OR UP TO 12 INCHES APART. ANOTHER PERSON WILL HOLD YOUR ANKLES WITH THE HANDS ONLY. NO OTHER METHOD OF BRACING OR HOLDING THE FEET IS AUTHORIZED. THE HEEL IS THE ONLY PART OF YOUR FOOT THAT MUST STAY IN CONTACT WITH THE GROUND. YOUR FINGERS MUST BE INTERLOCKED BEHIND YOUR HEAD AND THE BACKS OF YOUR HANDS MUST TOUCH THE GROUND. YOUR ARMS AND ELBOWS NEED NOT TOUCH THE GROUND. ON THE COMMAND "GO", BEGIN RAISING YOUR UPPER BODY FORWARD TO, OR BEYOND, THE VERTICAL POSITION. THE VERTICAL POSITION MEANS THAT THE BASE OF YOUR NECK IS ABOVE THE BASE OF YOUR SPINE. AFTER YOU HAVE REACHED OR SURPASSED THE VERTICAL POSITION, LOWER YOUR BODY UNTIL THE BOTTOM OF YOUR SHOULDER BLADES TOUCH THE GROUND. YOUR HEAD, HANDS, ARMS, OR ELBOWS DO NOT HAVE TO TOUCH THE GROUND. AT THE END OF EACH REPETITION, THE SCORER WILL STATE THE NUMBER OF SIT-UPS YOU HAVE CORRECTLY COMPLETED. A REPETITION WILL NOT COUNT IF YOU FAIL TO REACH THE VERTICAL POSITION, FAIL TO KEEP YOUR FINGERS INTERLOCKED BEHIND YOUR HEAD, ARCH OR BOW YOUR BACK AND RAISE YOUR BUTTOCKS OFF THE

ADDITIONAL POINTS TO DEMONSTRATE FOR THE SIT-UP EVENT

The following points must be clarified during the demonstration:

- To minimize stress to the neck, it is recommended that the soldier keep his chin curled downward and touching the top of his chest throughout the performance of the sit-up event.
- From the starting (down) position, or during any phase of the sit-up, the soldier may not use his hands or arms to pull himself up to push off the ground (floor or mat) in order to help himself attain the up position. Any of these procedures can give the violator an unfair advantage. They also violate the intent of the event. The sit-up event will be terminated immediately for those soldiers who, by pushing or pulling, use their arms to assist themselves in attaining the up position.
- If a mat is used, the entire body, including the feet and head, must be on the mat at the start.
- From the starting (down) position, or during any phase of the sit-up, the soldier may not swing his hands or arms in order to help himself attain the up position. If this occurs, that repetition does not count.
- The soldier may wiggle to attain the up position. This gives him no advantage.
- While in the up position, the soldier may not help himself stay in that position by using the elbows or any part of the arms to lock on to or brace against the legs. The elbows can go either inside or outside the knees. However, to push or pull them into the sides or tops of the knees to get extra leverage and rest gives an unfair advantage to that soldier. Therefore, soldiers who use this technique will be warned once for the first violation and immediately terminated if the violation continues or recurs.
- During the performance of the sit-up event, the fingers must be interlocked and behind the head. As long as any of the fingers are overlapping to any degree, the fingers are considered to be interlocked.
- If either foot breaks contact with the ground during a repetition, that repetition will not count. Both heels must stay in contact with the ground (floor or mat) during the performance of the event. The scorer should ensure that the holder has the soldier's feet properly secured. The scorer tells the soldier if his heel(s) is raised from the ground and that the repetition will not count.

Figure 14-6

GROUND TO RAISE YOUR UPPER BODY, OR LET YOUR KNEES EXCEED A 90-DEGREE ANGLE. IF A REPETITION DOES NOT COUNT, THE SCORER WILL REPEAT THE NUMBER OF YOUR LAST CORRECTLY PERFORMED SIT-UP. THE UP POSITION IS THE ONLY AUTHORIZED REST POSITION. IF YOU STOP AND REST IN THE DOWN (STARTING) POSITION, THE EVENT WILL BE TERMINATED. AS LONG AS YOU MAKE A CONTINUOUS PHYSICAL EFFORT TO SIT UP, THE EVENT WILL NOT BE TERMINATED. YOU MAY NOT USE YOUR HANDS OR ANY OTHER MEANS TO PULL OR PUSH YOURSELF UP TO THE UP (RESTING) POSITION OR TO HOLD YOURSELF IN THE REST POSITION. IF YOU DO SO, YOUR PERFORMANCE IN THE EVENT WILL BE TERMINATED. CORRECT PERFORMANCE IS IMPORTANT. YOU WILL HAVE TWO MINUTES TO PERFORM AS MANY SIT-UPS AS YOU CAN. WATCH THIS DEMONSTRATION." (The exercise is then demonstrated. See Figure 14-6 for a list of points that need to be made during the demonstration.) "WHAT ARE YOUR QUESTIONS?"

Administration

After reading the instructions, the supervisor answers questions. He then moves the groups to their testing stations. The event supervisor cannot be a scorer. At this point, the testing is ready to begin. Successive groups do the event until all soldiers have completed it.

Timing Techniques

The event supervisor is the timer. He calls out the time remaining every 30 seconds and every second for the last 10 seconds of the two minutes. He ends the event after two minutes by the command "Halt!"

Scorers' Duties

The scorer may either kneel or sit about three feet from the testee's hip. The scorer's head should be about even with the testee's shoulder when the latter is in the vertical (up) position. Each scorer decides for himself whether to sit or kneel down when scoring. He may not lie down or stand while scoring. The scorer counts aloud the number of correctly performed sit-ups and repeats the number of the last correctly performed repetition if an incorrect one is done. Scorers tell the testees what they are doing wrong as it occurs during the event. A critique of his performance is given to each soldier after the event. When the soldier completes the event, the scorer records the number of correctly performed sit-ups, initials the scorecard, and returns it to the soldier.

When checking for correct body position, the scorer must be sure that at a 90-degree angle is formed at each knee by the soldier's upper and lower leg. The angle to be measured is not the one formed by the lower leg and the ground. If, while performing the sit-up event, this angle becomes greater than 90 degrees, the scorer should instruct the testee and holder to reposition the legs to the proper angle and obtain compliance before allowing the testee's performance to continue. The loss of the proper angle does not terminate the testee's performance in the event. When the soldier comes to the vertical position, the scorer must be sure that the base of the soldier's neck is above or past the base of the spine. A soldier who simply touches his knees with his elbows may not come to a completely vertical position. The scorer must ensure that the holder uses only his hands to brace the exerciser's feet.

TWO-MILE RUN

This event tests cardiorespiratory (aerobic) endurance and the endurance of the leg muscles. (See Figure 14-7.)

Equipment

Two stopwatches for the event supervisor, one clipboard and pen for each scorer, copies of the event's instructions and standards, and numbers for the testees are needed.

Facilities

There must be a level area with no more than a three-degree slope on which a measured course has been marked. An oval-shaped track of known length may be used. If a road course is used, the start and finish and one-mile (half way) point must be clearly marked.

Personnel

One event supervisor and at least one scorer for every 15 runners are required.

Instructions

The event supervisor must read the following: "THE TWO-MILE RUN IS USED TO ASSESS YOUR AEROBIC FITNESS AND YOUR LEG MUSCLES' ENDURANCE. You MUST COMPLETE THE RUN WITHOUT ANY PHYSICAL HELP. AT THE START, ALL SOLDIERS WILL LINE UP BEHIND THE STARTING LINE. ON THE COMMAND 'GO,' THE CLOCK WILL START. YOU WILL BEGIN RUNNING AT YOUR OWN PACE. TO RUN THE REQUIRED TWO MILES, YOU MUST COMPLETE (describe the number of laps, start and finish points, and course layout). YOU ARE BEING TESTED ON YOUR ABILITY TO COMPLETE THE 2-MILE COURSE IN THE SHORTEST TIME POSSIBLE. ALTHOUGH WALKING IS AUTHORIZED, IT IS STRONGLY DISCOURAGED. IF YOU ARE PHYSICALLY HELPED IN ANY WAY (FOR EXAMPLE, PULLED, PUSHED, PICKED UP, AND/OR CARRIED) OR LEAVE THE DESIGNATED RUNNING COURSE FOR ANY

TWO-MILE RUN

Figure 14-7

REASON, YOU WILL BE DISQUALIFIED. (IT IS LEGAL TO PACE A SOLDIER DURING THE 2-MILE RUN. AS LONG AS THERE IS NO PHYSICAL CONTACT WITH THE PACED SOLDIER AND IT DOES NOT PHYSICALLY HINDER OTHER SOLDIERS TAKING THE TEST, THE PRACTICE OF RUNNING AHEAD OF, ALONG SIDE OF, OR BEHIND THE TESTED SOLDIER, WHILE SERVING AS A PACER, IS PERMITTED. CHEERING OR CALLING OUT THE ELAPSED TIME IS ALSO PERMITTED.) THE NUMBER ON YOUR CHEST IS FOR IDENTIFICATION. YOU MUST MAKE SURE IT IS VISIBLE AT ALL TIMES. TURN IN YOUR NUMBER WHEN YOU FINISH THE RUN. THEN, GO TO THE AREA DESIGNATED FOR THE COOL-DOWN AND STRETCH. DO NOT STAY NEAR THE SCORERS OR THE FINISH LINE AS THIS MAY INTERFERE WITH THE TESTING. WHAT ARE YOUR QUESTIONS ON THIS EVENT?"

Administration

After reading the instructions, the supervisor answers questions. He then organizes the soldiers into groups of no more than 10. The scorer for each group assigns a number to each soldier in the group. At the same time, the scorer collects the scorecards and records each soldier's number.

Timing Techniques

The event supervisor is the timer. He uses the commands "Get set" and "Go." Two stopwatches are used in case one fails. As the soldiers near the finish line, the event supervisor calls off the time in minutes and seconds (for example, "Fifteen-thirty, fifteen-thirty-one, fifteen-thirty -two," and so on).

Scorers' Duties

The scorers observe those runners in their groups, monitor their laps (if appropriate), and record their times as they cross the finish line. (It is often helpful to record the soldiers' numbers and times on a separate sheet of paper or card. This simplifies the recording of finish times when large groups of soldiers are simultaneously tested.) After all runners have completed the run, the scorers determine the point value for each soldier's run time, record the point values on the scorecards, and enter their initials in the scorers' blocks. In all cases, when a time falls between two point values, the lower point value is used and recorded. For example, if a female soldier, age 17 to 21, runs the two miles in 15 minutes and 19 seconds, the score awarded is 95 points.

At this time, the scorers for the 2-mile run also convert the raw scores for the push-up and sit-up events by using the scoring standards on the back side of the scorecard. They enter those point values on the scorecards and determine the total APFT score for each soldier before giving the scorecards to the test's OIC or NCOIC. After the test scores have been checked, the test's OIC or NCOIC signs all scorecards and returns them to the unit's commander or designated representative.

Test Results

The soldier's fitness performance for each APFT event is determined by converting the raw score for each event to a point score.

Properly interpreted, performance on the APFT shows the following:

- Each soldier's level of physical fitness.
- The entire unit's level of physical fitness.
- Deficiencies in physical fitness.
- Soldiers who need special attention.

(Leaders must develop special programs to improve the performance of soldiers who are below the required standards.)

Commanders should not try to determine the individual's or the unit's strengths and weaknesses in fitness by using only the total scores. A detailed study of the results on each event is more important. For a proper analysis of the unit's performance, event scores should be used. They are corrected for age and sex. Therefore, a female's 80-point push-up score should be considered the same as a male's 80-point push-up score. Using the total point value or raw scores may distort the interpretation.

Scores Above Maximum

Even though some soldiers exceed the maximum score on one or more APFT events, the official, maximum score on the APFT must remain at 300 (100 points per event). Some commanders, however, want to know unofficial point scores to reward soldiers for their extra effort.

Only those soldiers who score 100 points in all three events are eligible to determine their score on an extended scale. To fairly determine the points earned, extra points are awarded at the same rate as points obtained for scores at or below the 100 point level. Each push-up and sit-up beyond the maximum is worth one point as is every six-second decrease in the run time. Take, for example, the following case shown in Figure 14-8. A male soldier performs above the maximum in the 17-21 age group by doing 87 push-ups and 98 sit-ups and by running the two miles in 11 minutes and 12 seconds. His score would be calculated as follows:

CALCULATION FOR AGE 17-21 MALE

PUSH-UPS	
Actual	87
Maximum	82
Additional points	+5
Points (official)	100
Points (unofficial)	105
SIT-UPS	
Actual	98
Maximum	92
Additional points	+6
Points (official)	100
Points (unofficial)	106
2-MILE RUN	
Actual	11:12
Maximum	11:54
	:42
ADDITIONAL POINTS (42 sec/6 = 7)	+7
Points (official)	100
Points (unofficial)	107

Thus, the unofficial total for this soldier in the three events is determined in the following way:

PUSH-UPS	105
SIT-UPS	106
2-MILE RUN	+ 107
UNOFFICIAL TOTAL	318

Figure 14-8

The calculations on the previous page, give the soldier a total score of 318 points. This method lets the commander easily determine the scores for performances that are above the maximum. He may recognize soldiers for their outstanding fitness achievements, not only on the APFT but also for other, unofficial fitness challenges. Using this method ensures that each soldier has an equal chance to be recognized for any of the tested fitness components. Commanders may also establish their own incentive programs and set their own unit's standards (AR 350-15).

Temporary Profiles

A soldier with a temporary profile must take the regular three-event APFT after the profile has expired. (Soldiers with temporary profiles of more than three months may take an alternate test as determined by the commander with input from health-care personnel.) Once the profile is lifted, the soldier must be given twice the time of the profile (but not more than 90 days) to train for the APFT. For example, if the profile period was 7 days, the soldier has 14 days to train for the APFT after the profile period ends. If a normally scheduled APFT occurs during the profile period, the soldier should be given a mandatory make-up date.

Permanent Profiles

A permanently profiled soldier is given a physical training program by the profiling officer using the positive profile form DA 3349 (see Appendix B). The profiling officer gives the unit's commander a list of physical activities that are suitable for the profiled soldier. He also indicates the events and/or alternate aerobic event that the soldier will do on the APFT. This recommendation, made after consultation with the profiled soldier, should address the soldier's abilities and preference and the equipment available. (See DA Form 3349, Physical Profile, referenced in AR 40-501.)

The profiled soldier must perform all the regular APFT events his medical profile permits. Each soldier must earn at least 60 points on the regular events to receive a "go." He must also complete the alternate event in a time equal to or less than the one listed for his age group. For example, a soldier whose profile forbids only running will do the push-up and sit-up events and an alternate aerobic event. He must get at least a minimum passing score on each event to earn a "go" for the test. A soldier whose profile prevents two or more APFT events must complete the 2-mile run or an alternate aerobic event to earn a "go" on the test. Soldiers who cannot do any of the aerobic events due to a profile cannot be tested. Such information will be recorded in their official military record.

The standards for alternate events are listed in Figure 14-9. Scoring for all alternate events is on a go/no go basis. Soldiers who do push-up and sit-up events but who take an alternate aerobic event are not awarded promotion points for APFT performance.

Alternate Events

Alternate APFT events assess the aerobic fitness and muscular endurance of soldiers with permanent medical profiles or long-term (greater than three months) temporary profiles who cannot take the regular, three-event APFT.

The alternate aerobic APFT events are the following:

- 800-yard-swim test.
- 6.2-mile-stationary- bicycle ergometer test with a resistance setting of 2 kiloponds (2 kilograms) or 20 newtons.
- 6.2-mile-bicycle test on a conventional bicycle using one speed.
- 2.5-mile-walk test.

ALTERNATE TEST STANDARDS BY EVENT, SEX, AND AGE											
EVENT	SEX	AGE									
		17-21	22-26	27-31	32-36	37-41	42-46	47-51	52-56	57-61	62+
800-YARD SWIM	Men Women	20:00 21:00	20:30 21:30	21:00 22:00	21:30 22:30	22:00 23:00	22:30 23:30	23:00 24:00	24:00 25:00	24:30 25:30	25:00 26:00
6.2-MILE BIKE (Stationary and track)	Men Women	24:00 25:00	24:30 25:30	25:00 26:00	25:30 26:30	26:00 27:00	27:00 28:00	28:00 30:00	30:00 32:00	31:00 33:00	32:00 34:00
2.5-MILE WALK	Men Women	34:00 37:00	34:30 37:30	35:00 38:00	35:30 38:30	36:00 39:00	36:30 39:30	37:00 40:00	37:30 40:30	38:00 41:00	38:30 41:30

***Figure 14-9**

800-YARD-SWIM TEST

This event is used to assess cardiorespiratory (aerobic) fitness. (See Figure 14-10.)

Equipment

Two stopwatches, one clipboard and pen for each scorer, one copy each of the test instructions and standards, and appropriate safety equipment are needed.

Facilities

A swimming pool at least 25 yards long and 3 feet deep, or an approved facility, is needed.

Personnel

One event supervisor and at least one scorer for every soldier to be tested are required. Appropriate safety, control, and medical personnel must also be present.

Instructions

The event supervisor must read the following statement: "THE 800-YARD SWIM IS USED TO ASSESS YOUR LEVEL OF AEROBIC FITNESS. YOU WILL BEGIN IN THE WATER; NO DIVING IS ALLOWED. AT THE START, YOUR BODY MUST BE IN CONTACT WITH THE WALL OF THE POOL. ON THE COMMAND 'GO,' THE CLOCK WILL START. YOU SHOULD THEN BEGIN SWIMMING AT YOUR OWN PACE, USING ANY STROKE OR COMBINATION OF STROKES YOU WISH. YOU MUST SWIM (tell the number) LAPS TO COMPLETE THIS DISTANCE. YOU MUST TOUCH THE WALL OF THE POOL AT EACH END OF THE POOL AS YOU TURN. ANY TYPE OF TURN IS AUTHORIZED. YOU WILL BE SCORED ON YOUR ABILITY TO COMPLETE THE SWIM IN A TIME EQUAL TO, OR LESS THAN, THAT LISTED FOR YOUR AGE AND SEX. WALKING ON THE BOTTOM TO RECUPERATE IS AUTHORIZED. SWIMMING GOGGLES ARE PERMITTED, BUT NO OTHER EQUIPMENT IS AUTHORIZED. WHAT ARE YOUR QUESTIONS ABOUT THIS EVENT?"

Figure 14-10

Administration

After reading the instructions, the event supervisor answers only related questions. He assigns one soldier to each lane and tells the soldiers to enter the water. He gives them a short warm-up period to acclimate to the water temperature and loosen up. Above all, the event supervisor must be alert to the safety of the testees throughout the test.

Timing Techniques

The event supervisor is the timer. He uses the commands "Get set" and "Go." Two stopwatches are used in case one fails. As the soldiers near the finish, the event supervisor begins calling off the elapsed time in minutes and seconds (for example, "Nineteen-eleven, nineteen-twelve, nineteen-thirteen," and so on). The time is recorded when each soldier touches the end of the pool on the final lap or crosses a line set as the 800-yard mark.

Scorers' Duties

Scorers must observe the swimmers assigned to the. They must be sure that each swimmer touches the bulkhead at every turn. The scorers record each soldier's time in the 2-mile-run block on the scorecard and use the comment block to identify the time as an 800-yard-swim time. If the pool length is measured in meters, the scorers convert the exact distance to yards. To convert meters to yards, multiply the number of meters by 39.37 and divide the product by 36; that is, (meters x 39.37)/36 = yards. For example, 400 meters equals 437.4 yards; that is, (400 x 39.37)/36 = 437.4 yards.

6.2-MILE STATIONARY-BICYCLE ERGOMETER TEST

This event is used to assess the soldier's cardiorespiratory and leg-muscle endurance. (See Figure 14-11.)

Equipment

Two stopwatches, one clipboard and pen for each scorer, a copy of the test instructions and standards, and one stationary bicycle ergometer are needed. The ergometers should measure resistance in kiloponds or newtons. The bicycle should be one that can be used for training and testing. Its seat and

6.2 MILE STATIONARY-BICYCLE ERGOMETER TEST

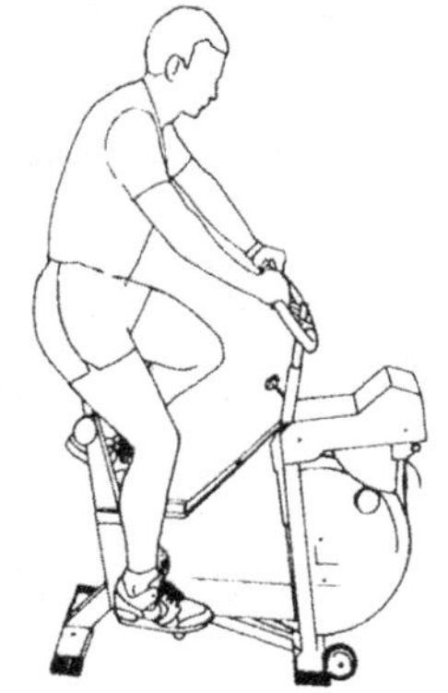

Figure 14-11

handlebars must be adjustable to let the soldier fully extend his legs when pedaling. It should have an adjustable tension setting and an odometer. The resistance is usually set by a tension strap on a weighted pendulum connected to the flywheel. See Appendix D for guidance on using various types of stationary bikes.

Facilities

The test site can be any place where there is an approved bicycle ergometer. This could be the post's fitness facility or the hospital's therapy clinic. Each test station must be two yards wide and four yards deep.

Personnel

One event supervisor and at least one scorer for every three soldiers to be tested are required. Appropriate safety, control, and medical personnel should also be present.

Instructions

The event supervisor must read the following: "THE 6.2-MILE STATIONARY-BICYCLE ERGOMETER EVENT TESTS YOUR CARDIORESPIRATORY FITNESS AND LEG MUSCLE ENDURANCE. THE ERGOMETER'S RESISTANCE MUST BE SET AT TWO KILOPOUNDS (20 NEWTONS). ON THE COMMAND 'GO,' THE CLOCK WILL START, AND YOU WILL BEGIN PEDALING AT YOUR OWN PACE WHILE MAINTAINING THE RESISTANCE INDICATOR AT TWO POUNDS. YOU WILL BE SCORED ON YOUR ABILITY TO COMPLETE 6.2 MILES (10 KILOMETERS), AS SHOWN ON THE ODOMETER, IN A TIME EQUAL TO OR LESS THAN THAT LISTED FOR YOUR AGE AND SEX. WHAT ARE YOUR QUESTIONS ABOUT THIS EVENT?"

Administration

After reading the instructions, the event supervisor answers any related questions. Each soldier is given a short warm-up period and allowed to adjust the seat and handlebar height.

Timing Techniques

The event supervisor is the timer. He uses the commands "Get set" and "Go." Two stopwatches are used in case one fails. As the soldiers pedal the last two-tenths of the test distance, the event supervisor should start calling off the time in minutes and seconds (for example, "Twenty-thirty-one, twenty -thirty -two, twenty-thirty-three," and so on). He calls the time remaining every 30 seconds for the last two minutes of the allowable time and every second during the last ten seconds.

Scorers' Duties

Scorers must ensure that the bicycle ergometer is functioning properly. They must then make sure that the bicycle ergometers' tension settings have been calibrated and are accurate and that the resistance of the ergometers has been set at two kiloponds (20 newtons). The scorers must observe the soldiers throughout the event. From time to time the scorer may need to make small adjustments to the resistance control to ensure that a continuous resistance of exactly 2 kiloponds (20 newtons) is maintained throughout the test. At the end of the test, they record each soldier's time on the scorecard in the 2-mile-run block, initial the appropriate block, and note in the comment block that the time is for a 6.2-mile stationary-bicycle ergometer test.

6.2-MILE BICYCLE TEST

This event is used to assess the soldier's cardiorespiratory and leg-muscle endurance.

Equipment

Two stopwatches, one clipboard and pen for each scorer, a copy of the test instructions and standards, and numbers are needed. Although one-speed bicycles are preferred for this event, multispeed bicycles may be used. If a multispeed bicycle is used, measures must be taken to ensure that only one gear is used throughout the test. (This can usually be done by taping the gear shifters at the setting preferred by the testee.)

Facilities

A relatively flat course with a uniform surface and no obstacles must be used. It must also be clearly marked. Soldiers should not be tested on a quarter-mile track, and they should never be out of the scorers' sight. The course should be completely free of runners and walkers.

Personnel

One event supervisor and at least one scorer for every 10 soldiers are required. Safety, control, and medical personnel should also be present as appropriate.

instructions

The event supervisor must read the following: "THE 6.2-MILE BICYCLE TEST IS USED TO ASSESS YOUR CARDIORESPIRATORY FITNESS AND LEG MUSCLES' ENDURANCE. YOU MUST COMPLETE THE 6.2

MILES WITHOUT ANY PHYSICAL HELP FROM OTHERS. YOU MUST KEEP YOUR BICYCLE IN ONE GEAR OF YOUR CHOOSING FOR THE ENTIRE TEST. CHANGING GEARS IS NOT PERMITTED AND WILL RESULT IN DISQUALIFICATION. TO BEGIN, YOU WILL LINE UP BEHIND THE STARTING LINE. ON THE COMMAND 'GO,' THE CLOCK WILL START, AND YOU WILL BEGIN PEDALING AT YOUR OWN PACE. TO COMPLETE THE REQUIRED DISTANCE OF 6.2 MILES, YOU MUST COMPLETE (describe the number of laps, start and finish points, and course layout). YOU WILL BE SCORED ON YOUR ABILITY TO COMPLETE THE DISTANCE OF 6.2 MILES (10 KILOMETERS) IN A TIME EQUAL TO OR LESS THAN THAT LISTED FOR YOUR AGE AND SEX. IF YOU LEAVE THE DESIGNATED COURSE FOR ANY REASON, YOU WILL BE DISQUALIFIED. WHAT ARE YOUR QUESTIONS ABOUT THIS EVENT?"

Administration

After reading the instructions, the event supervisor answers any related questions. He then organizes the soldiers into groups of no more than ten and assigns each group to a scorer. Scorers assign numbers to the soldiers in their groups and record each soldier's number on the appropriate scorecard.

Timing Techniques

The event supervisor is the timer. He uses the commands "Get set" and "Go." Two stopwatches are used in case one fails. As soldiers near the end of the 6.2-mile ride, the event supervisor starts calling off the time in

6.2-MILE BICYCLE TEST

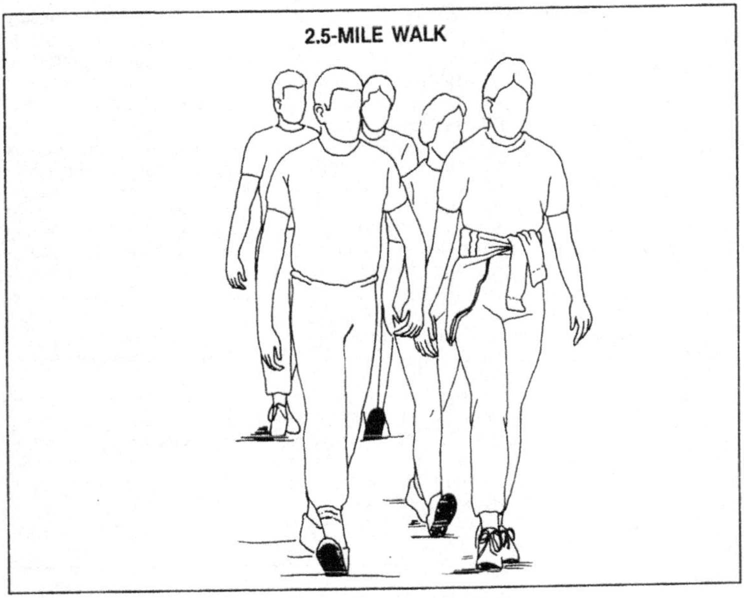

minutes and seconds (for example, "Thirty-twenty-one, thirty- twenty-two, thirty -twenty-three," and so on).

Scorers' Duties

When the event is over, scorers record each soldier's time in the 2-mile-run block. They initial the appropriate block and note in the comment block that the time is for a 6.2-mile-bicycle test and whether or not the testee met the required standards for his age and sex.

2.5-MILE WALK

This event serves to assess cardiorespiratory and leg-muscle endurance.

Equipment

Two stopwatches, one clipboard and pen for each scorer, numbers, and copies of the test instructions and standards are needed.

Facilities

This event uses the same course as the 2-mile run.

Personnel

One event supervisor and at least one scorer for every three soldiers to be tested are required. Appropriate safety, control, and medical personnel should be present.

Instructions

The event supervisor must read the following: "THE 2.5-MILE WALK IS USED TO ASSESS YOUR CARDIORESPIRATORY FITNESS AND LEG-MUSCLE ENDURANCE. ON THE COMMAND 'GO,' THE CLOCK WILL START, AND YOU WILL BEGIN WALKING AT YOUR OWN PACE. YOU MUST COMPLETE (describe the number of laps, start and finish points, and course layout). ONE FOOT MUST BE IN CONTACT WITH THE GROUND AT ALL TIMES. IF YOU BREAK INTO A RUNNING STRIDE AT ANY TIME OR HAVE BOTH FEET OFF THE GROUND AT THE SAME TIME, YOUR PERFORMANCE IN THE EVENT WILL BE TERMINATED. YOU WILL BE SCORED ON YOUR ABILITY TO COMPLETE THE 2.5-MILE COURSE IN A TIME EQUAL TO OR LESS THAN THAT LISTED FOR YOUR AGE AND SEX. WHAT ARE YOUR QUESTIONS ABOUT THIS EVENT?"

Administration

After reading the instructions, the event supervisor answers any related questions. He then divides the soldiers into groups of no more than three and assigns each group to a scorer. Each soldier is issued a number which the scorer records on the scorecard.

Timing Techniques

The event supervisor is the timer. He uses the commands "Get set" and "Go." Two stopwatches are used in case one fails. As the soldiers near the end of the 2.5-mile walk, the event supervisor starts calling off the elapsed time in minutes and seconds (for example,"Thirty-three-twenty-two, thirty -three -twenty -three, thirty-three-twenty -four," and so on).

Scorers' Duties

Scorers must observe the soldiers during the entire event and must ensure that the soldiers maintain a walking stride. Soldiers who break into any type of running stride will be terminated from the event and given a "no go." When the event is over, scorers record the time in the 2-mile-run block on the scorecard, initial the appropriate block, and note in the comment block that the time is for a 2.5-mile walk and whether or not the testee received a "go" or "no go."

APPENDIX A

PHYSIOLOGICAL DIFFERENCES BETWEEN THE SEXES

Soldiers vary in their physical makeup. Each body reacts differently to varying degrees of physical stress, and no two bodies react exactly the same way to the same physical stress. For everyone to get the maximum benefit from training, leaders must be aware of these differences and plan the training to provide maximum benefit for everyone. They must also be aware of the physiological differences between men and women. While leaders must require equal efforts of men and women during the training period, they must also realize that women have physiological limitations which generally preclude equal performance. The following paragraphs describe the most important physical and physiological differences between men and women.

SIZE

The average 18- year-old man is 70.2 inches tall and weighs 144.8 pounds, whereas the average woman of the same age is 64.4 inches tall and weighs 126.6 pounds. This difference in size affects the absolute amount of physical work that can be performed by men and women.

MUSCLES

Men have 50 percent greater total muscle mass, based on weight, than do women. A woman who is the same size as her male counterpart is generally only 80 percent as strong. Therefore, men usually have an advantage in strength, speed, and power over women.

FAT

Women carry about 10 percentage points more body fat than do men of the same age. Men accumulate fat primarily in the back, chest, and abdomen; women gain fat in the buttocks, arms, and thighs. Also, because the center of gravity is lower in women than in men, women must overcome more resistance in activities that require movement of the lower body.

BONES

Women have less bone mass than men, but their pelvic structure is wider. This difference gives men an advantage in running efficiency.

HEART SIZE AND RATE

The average woman's heart is 25 percent smaller than the average man's. Thus, the man's heart can pump more blood with each beat. The larger heart size contributes to the slower resting heart rate (five to eight beats a minute slower) in males. This lower rate is evident both at rest and at any given level of submaximal exercise. Thus, for any given work rate, the faster heart rate means that most women will become fatigued sooner than men.

FLEXIBILITY

Women generally are more flexible than men.

LUNGS

The lung capacity of men is 25 to 30 percent greater than that of women. This gives men still another advantage in the processing of oxygen and in doing aerobic work such as running.

RESPONSE TO HEAT

A woman's response to heat stress differs somewhat from a man's. Women sweat less, lose less heat through evaporation, and reach higher body temperatures before sweating starts. Nevertheless, women can adapt to heat stress as well as men. Regardless of gender, soldiers with a higher level of physical fitness generally better tolerate, and adapt more readily to, heat stress than do less fit soldiers.

OTHER FACTORS

Knowing the physiological differences between men and women is just the first step in planning physical training for a unit. Leaders need to understand other factors too.

Women can exercise during menstruation; it is, in fact, encouraged. However, any unusual discomfort, cramps, or pains while menstruating should be medically evaluated.

Pregnant soldiers cannot be required to exercise without a doctor's approval. Generally, pregnant women may exercise until they are close to childbirth if they follow their doctors' instructions. The Army agrees with the position of the American College of obstetricians and Gynecologists regarding exercise and pregnancy. This guidance is available from medical authorities and the U.S. Army Physical Fitness School (USAPFS). The safety and health of the mother and fetus are primary concerns when dealing with exercise programs.

Vigorous activity does not harm women's reproductive organs or cause menstrual problems. Also, physical fitness training need not damage the breasts. Properly fitted and adjusted bras, however, should be worn to avoid potential injury to unsupported breast tissue that may result from prolonged jarring during exercise.

Although female soldiers must sometimes be treated differently from males, women can reach high levels of physical performance. Leaders must use common sense to help both male and female soldiers achieve acceptable levels of fitness. For example, ability-group running alleviates gender-based differences between men and women. Unit runs, however, do not.

APPENDIX B

POSITIVE PROFILE FORM

PHYSICAL PROFILE

For use of this form, see AR 40-501; the proponent agency is the Office of The Surgeon General

1 MEDICAL CONDITION

2

P	U	L	H	E	S

3 ASSIGNMENT LIMITATIONS ARE AS FOLLOWS

4 THIS PROFILE IS ☐ PERMANENT ☐ TEMPORARY EXPIRATION DATE

5 THE ABOVE STATED MEDICAL CONDITION SHOULD NOT PREVENT THE INDIVIDUAL FROM DOING THE FOLLOWING ACTIVITIES

☐ GROIN STRETCH	☐ THIGH STRETCH	☐ LOWER BACK STRETCH	☐ NECK & SHLDR STRETCH	☐ NECK STRETCH
☐ HIP RAISE	☐ QUADS STRETCH & BAL	☐ SINGLE KNEE TO CHEST	☐ UPPER BACK STRETCH	☐ ANKLE STRETCH
☐ KNEE BENDER	☐ CALF STRETCH	☐ STRAIGHT LEG RAISE	☐ CHEST STRETCH	☐ HIP STRETCH
☐ SIDE STRADDLE HOP	☐ LONG SIT	☐ ELONGATION STRETCH	☐ ONE ARM SIDE STRETCH	☐ UPPER BODY WT TNG
☐ HIGH JUMPER	☐ HAMSTRING STRETCH	☐ TURN AND BOUNCE	☐ TWO ARM SIDE STRETCH	☐ LOWER BODY WT TNG
☐ JOGGING IN PLACE	☐ HAMS & CALF STRETCH	☐ TURN AND BEND	☐ SIDE BENDER	☐ ALL

6 AEROBIC CONDITIONING EXERCISES

- ☐ WALK AT OWN PACE AND DISTANCE
- ☐ RUN AT OWN PACE AND DISTANCE
- ☐ BICYCLE AT OWN PACE AND DISTANCE
- ☐ SWIM AT OWN PACE AND DISTANCE
- ☐ WALK OR RUN IN POOL AT OWN PACE
- ☐ UNLIMITED WALKING
- ☐ UNLIMITED RUNNING
- ☐ UNLIMITED BICYCLING
- ☐ UNLIMITED SWIMMING
- ☐ RUN AT TRAINING HEART RATE FOR ____ MIN
- ☐ BICYCLE AT TRAINING HEART RATE FOR ____ MIN
- ☐ SWIM AT TRAINING HEART RATE FOR ____ MIN

7 FUNCTIONAL ACTIVITIES

- ☐ WEAR BACKPACK (40 LBS)
- ☐ WEAR HELMET
- ☐ CARRY RIFLE
- ☐ FIRE RIFLE WITH HEARING PROTECTION
- ☐ KP/MOPPING/MOWING GRASS
- ☐ MARCHING UP TO ____ MILES
- ☐ LIFT UP TO ____ POUNDS
- ☐ ALL

FITNESS TEST

- ☐ TWO MILE RUN ☐ WALK
- ☐ PUSH UPS ☐ SWIM
- ☐ SIT UPS ☐ BICYCLE

8 TRAINING HEART RATE FORMULA

MALES 220 FEMALES 225

MINUS (-) AGE

MINUS (-) RESTING HEART RATE

TIMES (X) % INTENSITY

PLUS (+) RESTING HEART RATE

50% - EXTREMELY POOR CONDITION

60% HEALTHY, SEDENTARY INDIVIDUAL

70% MODERATELY ACTIVE, MAINTENANCE

80% WELL TRAINED PERSON

9 OTHER

SAMPLE

TYPED NAME AND GRADE OF PROFILING OFFICER	SIGNATURE	DATE
TYPED NAME AND GRADE OF PROFILING OFFICER	SIGNATURE	

ACTION BY APPROVING AUTHORITY

PERMANENT CHANGE OF PROFILE IS ☐ APPROVED ☐ NOT APPROVED

TYPED NAME, GRADE, & TITLE OF APPROVING AUTHORITY	SIGNATURE	DATE

ACTION BY UNIT COMMANDER

THIS PERMANENT CHANGE IN THE PHYSICAL PROFILE SERIAL ☐ DOES ☐ DOES NOT REQUIRE A CHANGE IN THE MEMBER'S ☐ MILITARY OCCUPATIONAL SPECIALTY ☐ DUTY ASSIGNMENT BECAUSE

TYPED NAME AND GRADE OF UNIT COMMANDER	SIGNATURE	DATE

PATIENT'S IDENTIFICATION (For typed or written entries give: Name - Last, first, middle; grade; date; hospital or medical facility)

ISSUING CLINIC AND PHONE NUMBER

DISTRIBUTION

UNIT COMMANDER – ORIGINAL & 1 COPY
HEALTH RECORD JACKET - 1 COPY
CLINIC FILE - 1 COPY
HQDA (DAPC EPA) 2461 EISENHOWER AVE ALEXANDRIA, VA 20310-2200 - 1 COPY

DA FORM 3349, MAY 86 REPLACES DA FORM 5302-R (TEST) AND DA FORM 3349 DATED 1 JUN 80.

Figure B-1

APPENDIX C

PHYSICAL FITNESS LOG

Soldiers can use a physical fitness log to record their fitness goals. The log will serve as a diary of how well they achieve them. Fitness goals are determined before the training begins. The results should closely parallel or exceed the unit's goals. While this is not a requirement, the log may also be used by commanders and supervisors as a record of physical fitness training. Figure C-1 shows an example of a physical fitness log that could be reproduced locally.

PHYSICAL FITNESS LOG

Name (Last) First MI	Rank	Sex (Circle) M F	Organization

Week	Date	Phase of Training	Exercise Activity No. 1	Reps, Time, Distance, Etc.	Heart Rate Attained	Exercise Activity No. 2	Reps, Time, Distance, Etc.	Heart Rate Attained

SAMPLE

INSTRUCTIONS: 1. Record each workout by entry of date, phase of training, activities performed, and heart rate attained. Also indicate the repetitions, time, or other indication of performance for each activity in the block provided.

2. At the end of each week, extend the last horizontal line to the left to indicate the end of the week, and place the number of the week in the weekly column.

Figure C-1

APPENDIX D

STATIONARY BICYCLE TEST

Only stationary bicycles which can be calibrated and which have mechanically adjustable resistances may be used to test profiled soldiers on the 6.2-mile (1 O-kilometer), alternate APFT event. Therefore, the event supervisor or scorer must be sure that the stationary bicycle can be accurately adjusted to ensure that the soldier pedals against the correct resistance (force) of 2 kiloponds or 20 newtons. If the stationary bicycle cannot be properly calibrated and adjusted, the soldier may end up pedalling against a resistance which is too great or not great enough. In either case, the test would not provide an accurate indication of the soldier's level of cardiorespiratory fitness.

The best type of stationary bicycle for testing has the following features:

- Calibration adjustment.
- Adjustable resistance displayed in kiloponds or newtons.
- Odometer which accurately measures the distance traveled in either miles or tenths of miles or in kilometers and tenths of kilometers.

Examples of stationary bicycles which meet the above criteria are the mechanically braked Bodyguard 990 and Monark 868. Such bicycles can be used to accurately measure a person's rate of work or the total amount of work. They are often called bicycle ergometers.

If the stationary bicycle has an odometer, the soldier must pedal 6.2 miles (10.0 kilometers or 10,000 meters) against a resistance set at 2 kiloponds or 20 newtons. The test is completed when the soldier pedals 6.2 miles (10.0 kilometers). He receives a "Go" if he is below or at the time allotted for his particular age group and gender. Care should be taken to ensure that, when using a stationary bicycle which measures distance in kilometers, the test is ended at 10 kilometers, not 6.2 kilometers.

There are many electrically operated, stationary bicycles (EOSBS) on the market and in gymnasiums on Army installations. Most of them are designed for physical fitness training. Only a limited number of EOSB models are designed to accurately assess a person's energy expenditure during exercise. Such EOSBS are relatively expensive and are generally found in medical and scientific laboratories. Very few, if any, are found in gymnasiums on Army installations.

Because most of the more common training EOSBS were not designed to accurately assess energy expenditure, they should not be used for the alternate, cardiorespiratory APFT event.

For the sake of accuracy and ease of administration, soldiers designated to be tested on either of the two bicycle protocols should be tested using a moving bicycle IAW the guidelines provided elsewhere in this field manuel. If the mechanical y- braked Bodyguard 990 or Monark 868 is used, however, the tester must ensure that the equipment has been properly calibrated prior to each test.

TABLE D-1								
MALES								
AGE (YEARS)	17-21	22-26	27-31	32-36	37-41	42-46	47-51	52+
TIME ALLOTTED (MINUTES)	24.0	24.5	25.0	25.5	26.0	27.0	28.0	30.0
CALORIES/MIN.	9.8	9.7	9.5	9.3	9.2	8.9	8.6	8.2
CALORIES/HR.	590	580	570	560	550	535	520	490
N.-METERS/SEC. or WATTS	139	136	133	131	128	124	119	111
TOTAL CALORIES EXPENDED	236	237	238	239	240	241	242	245

TABLE D-2								
FEMALES								
AGE YEARS	17-21	22-26	27-31	32-36	37-41	42-46	47-51	52+
TIME ALLOTTED (MINUTES)	25.0	25.5	26.0	26.5	27.0	28.0	30.0	32.0
CALORIES/MIN.	9.5	9.3	9.2	9.0	8.9	8.6	8.2	7.8
CALORIES/HR.	570	560	550	545	535	520	490	465
N.-METERS/SEC. or WATTS	133	131	128	126	124	120	110	104
TOTAL CALORIES	237.5	238	239	240	240.5	242	245	248

APPENDIX E

SELECTING THE RIGHT RUNNING SHOE

Choosing a running shoe that is suitable for your particular type of foot can help you avoid some common running- related injuries. It can also make running more enjoyable and let you get more mileage out of your shoes.

Shoe manufacturers are aware that, anatomically, feet usually fall into one of three categories. Some people have "floppy" feet that are very "loose- jointed." Because feet like this are too mobile, they "give" when they hit the ground. These people need shoes that are built to control the foot's motion. At the other extreme are people with "rigid" feet. These feet are very tight-jointed and do not yield enough upon impact. To help avoid impact-related injuries, these people need shoes that cushion the impact of running. Finally, the third type, or normal foot, falls somewhere between mobile and rigid. This type of foot can use any running shoe that is stable and properly cushioned. Use the chart at Figure E-1 to help you determine what kind of foot you have. Then, read the information on special features you should look for in a shoe.

When shopping for running shoes, keep the following in mind:

- Expect to spend between $30 and $100 for a pair of good shoes.
- Discuss your foot type, foot problems, and shoe needs with a knowledgeable salesperson.
- Check the PX for available brands and their prices before shopping at other stores.
- Buy a training shoe, not a racing shoe.
- When trying on shoes, wear socks that are as similar as possible to those in which you will run. Also, be sure to try on both shoes.
- Look at more than one model of shoe.
- Choose a pair of shoes that fit both feet well while you are standing.
- Ask if you can try running in the shoes on a non-carpeted surface. This gives you a feel for the shoes.
- Carefully inspect the shoes for defects that might have been missed by quality control. Do the following:
  - -Place the shoes on a flat surface and check the heel from behind to see that the heel cup is perpendicular to the sole of the shoe.
  - -Feel the seams inside the shoe to determine if they are smooth, even, and well-stitched.
  - -Check for loose threads or extra glue spots; they are usually signs of poor construction.

The shoes' ability to protect you from injury decreases as the mileage on them increases. Record the number of miles you run with them on a regular basis, and replace the shoes when they have accumulated 500 to 700 miles even if they show little wear.

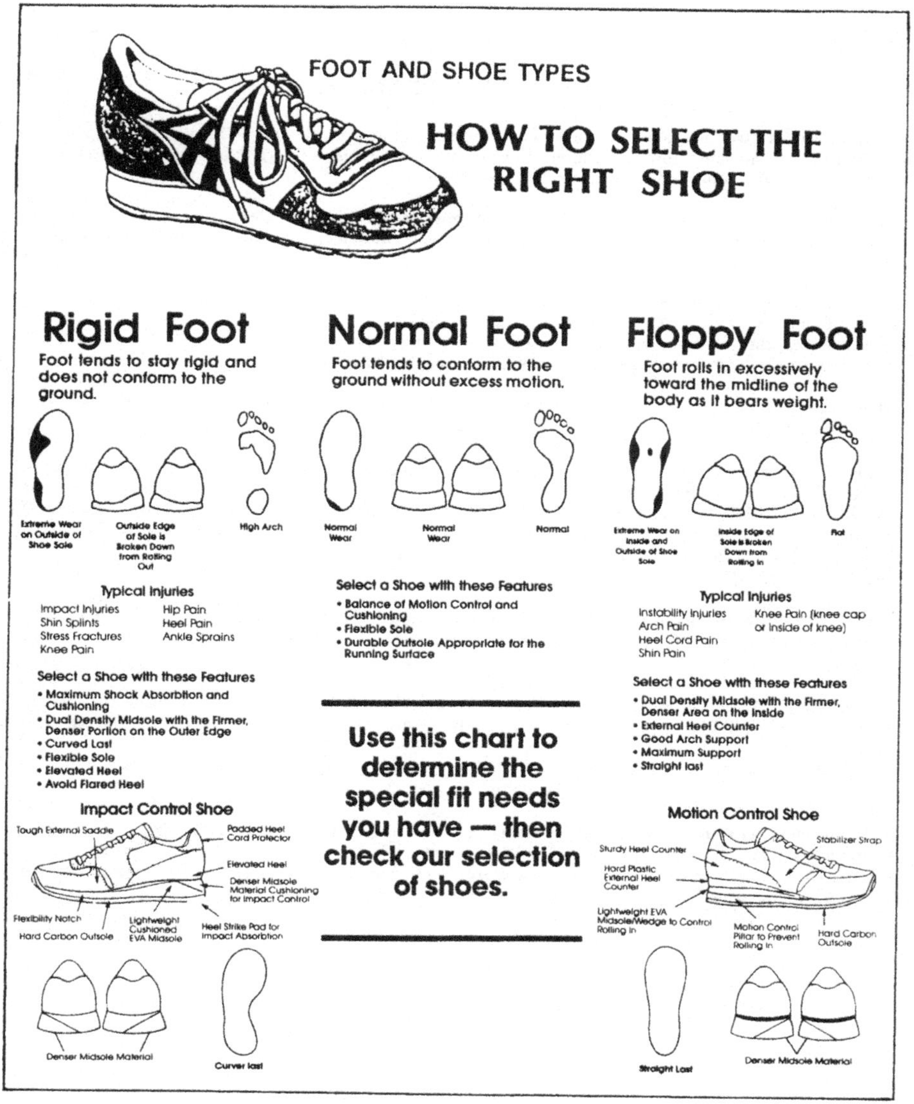

FOOT AND SHOE TYPES
HOW TO SELECT THE RIGHT SHOE
Rigid Foot
Foot tends to stay rigid and does not conform to the ground.
Extreme Wear on Outside of Shoe Sole
Outside Edge of Sole is Broken Down from Rolling Out
High Arch
Typical Injuries
Impact Injuries
Shin Splints
Stress Fractures
Knee Pain
Hip Pain
Heel Pain
Ankle Sprains
Select a Shoe with these Features
• Maximum Shock Absorbtion and Cushioning
• Dual Density Midsole with the Firmer, Denser Portion on the Outer Edge
• Curved Last
• Flexible Sole
• Elevated Heel
• Avoid Flared Heel
Impact Control Shoe
Tough External Saddle
Padded Heel Cord Protector
Elevated Heel
Denser Midsole Material Cushioning for Impact Control
Flexibility Notch
Hard Carbon Outsole
Lightweight Cushioned EVA Midsole
Heel Strike Pad for Impact Absorbtion
Denser Midsole Material
Curver last
Normal Foot
Foot tends to conform to the ground without excess motion.
Normal Wear
Normal Wear
Normal
Select a Shoe with these Features
• Balance of Motion Control and Cushioning
• Flexible Sole
• Durable Outsole Appropriate for the Running Surface
Use this chart to determine the special fit needs you have — then check our selection of shoes.
Floppy Foot
Foot rolls in excessively toward the midline of the body as it bears weight.
Extreme Wear on Inside and Outside of Shoe Sole
Inside Edge of Sole is Broken Down from Rolling In
Flat
Typical Injuries
Instability Injuries
Arch Pain
Heel Cord Pain
Shin Pain
Knee Pain (knee cap or inside of knee)
Select a Shoe with these Features
• Dual Density Midsole with the Firmer, Denser Area on the Inside
• External Heel Counter
• Good Arch Support
• Maximum Support
• Straight last
Motion Control Shoe
Sturdy Heel Counter
Stabilizer Strap
Hard Plastic External Heel Counter
Lightweight EVA Midsole/Wedge to Control Rolling In
Motion Control Pillar to Prevent Rolling In
Hard Carbon Outsole
Straight Last
Denser Midsole Material

Figure E-1

APPENDIX F

CALCULATION OF $\dot{V}O_2max$

This appendix gives a step-by-step example of how a soldier can calculate $\dot{V}O_2max$ using his all-out, 2-mile-run time. This lets interested soldiers compare their fitness levels with others such as athletes whose VO_2max values are published in magazines or journals.

The two equations below convert the 2-mile-run times of males and females to maximum oxygen uptake values. The VO_2max values obtained are shown as the maximum amount of oxygen in milliliters used per kilogram of the person's body weight in one minute during maximum aerobic exercise. $\dot{V}O_2max$ values are generally expressed more succinctly as ml O_2/kg x min.

For males, the following equation is used to calculate $\dot{V}O_2max$:

$\dot{V}O_2max$ = 99.7 - [3.35 x (2-mile-run time in decimal form)].

For females, the following equation is used:

$\dot{V}O_2max$ = 72.9 - [1.77 x (2-mile-run time in decimal form)].

The example below shows how to use the equation for males. The data is for a 21-year-old male whose all-out, 2-mile-run time is 12 minutes and 36 seconds.

STEP 1. Express the 2-mile-run time as a decimal, and insert it into the equation. When 12 minutes 36 seconds is written as a decimal, it becomes 12.60 minutes. (To determine what fraction of a minute 36 seconds is, divide 36 seconds by the number of seconds in one minute, that is, 60 seconds. Thus, 36/60 = 0.60. This fraction is added to the minute value to give 12.60 as the run time expressed as a decimal.) After putting the decimal form into the equation, the equation should resemble the one below.

$\dot{V}O_2max$ = 99.7 -[3.35 x (12.60)]

STEP 2. Multiply the decimal form of the 2-mile-run time by 3.35. In this case, we get [3.35 x (12.60)], which equals 42.21. At this point, the equation should resemble the one below.

VO_2max = 99.7 -[42.21].

STEP 3. Subtract the product obtained in Step 2 from 99.7. For our example, we make the following subtraction: 99.7 -[42.21]. This gives a value of 57.49. Thus, the equation should look like the one below.

$\dot{V}O_2max$ = 57.49

This calculation reveals that a male whose all-out, 2-mile-run time is 12 minutes 36 seconds will have a $\dot{V}O_2max$ of approximately 57.49 ml O_2/kg x min.

To determine how this value or others translates into fitness ratings, refer to Table F-1. It presents information for finding one's level of CR fitness based on $\dot{V}O_2max$. By matching a soldier's value for maximum oxygen uptake with those in the table corresponding to his age group and sex, one gets an adjectival rating (fair, good, superior, etc.).

$\dot{V}O_2max$ AND CR FITNESS CLASSIFICATIONS

CATEGORY	SEX	AGE 20-29	30-39	40-49	50-59	60+
SUPERIOR	MALE	54.0+*	52.5+	50.4+	47.1+	45.2+
	FEMALE	46.8+	43.9+	41.0+	36.8+	37.5+
EXCELLENT	MALE	48.2-51.4	46.8-50.4	44.1-48.2	41.0-45.3	38.1-42.5
	FEMALE	41.0-44.2	38.5-41.0	36.3-39.5	32.1-35.2	31.2-35.2
GOOD	MALE	44.2-47.0	42.4-45.3	39.9-43.9	36.7-39.5	33.6-36.7
	FEMALE	36.7-39.5	34.6-37.4	32.3-35.1	29.4-39.9	27.2-30.9
FAIR	MALE	41.0-43.9	38.9-41.6	36.7-39.5	33.8-36.1	30.2-32.4
	FEMALE	33.8-36.1	32.1-33.9	29.5-31.6	26.9-28.7	24.5-26.5
POOR	MALE	37.1-40.3	35.4-38.1	33.0-35.6	30.2-32.5	26.5-29.4
	FEMALE	30.6-32.7	28.7-31.9	26.5-29.4	24.3-26.1	22.8-24.0
VERY POOR	MALE	27.1-36.7	26.5-34.0	24.2-32.3	22.1-29.4	18.3-25.1
	FEMALE	22.6-29.4	22.5-28.0	20.8-25.6	21.1-23.7	17.9-22.1

*$\dot{V}O_2max$ is expressed in ml O_2/kg x min.

Table F-1

Table F-1 lists some values for $\dot{V}O_2max$ along with their associated CR fitness levels. This table was obtained from the Institute for Aerobic Research in Dallas, Texas. These values can be used to classify a soldier's level of CR fitness based on his $\dot{V}O_2max$.

APPENDIX G

PERCEIVED EXERTION

The heart rate has traditionally been used to estimate exercise intensity. However, evidence shows that a person's own perception of the intensity of his exercise can often be just as accurate as the heart rate in gauging his exercise intensity.

The scale in Figure G-1 lets a soldier rate his degree of perceived exertion (PE). This scale consists of numerical ratings for physical exercise followed by their associated descriptive ratings.

PERCEIVED EXERTION (PE) SCALE

NUMERICAL RATING	VERBAL RATING
6	
7	very, very light
8	
9	very light
10	
11	fairly light
12	
13	somewhat hard
14	
15	hard
16	
17	very hard
18	
19	very, very hard
20	

Figure G-1

To judge perceived exertion, estimate how difficult it feels to do the exercise. Do not be concerned with any one single factor such as shortness of breath or work intensity. Instead, try to concentrate on the total inner feeling of exertion.

Multiplying the rating of perceived exertion by 10 roughly approximates the heart rate during exercise. For example, a PE of 14, when multiplied by 10, equals 140.

Most soldiers with THRs between 130 and 170 BPM would exercise between a PE of 13 (somewhat hard) and 17 (very hard).

Although either percent of maximum heart rate or perceived exertion may be used during exercise, the most valid method for calculating THR is percent HRR.

THE MAJOR SKELETAL MUSCLES OF THE HUMAN BODY

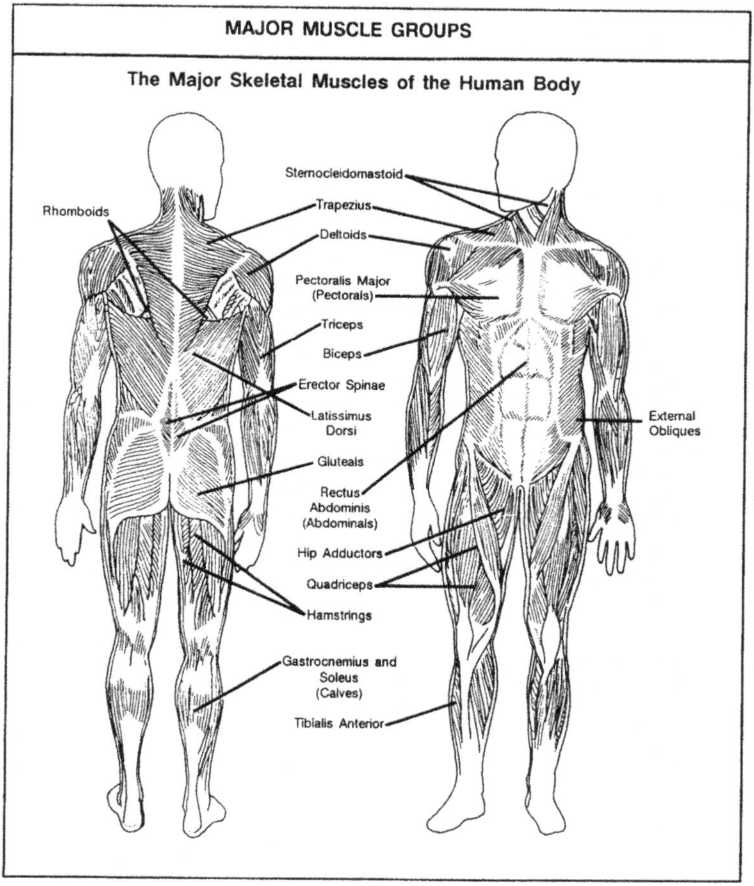

Figure H-1

The iliopoas muscle (a hip flexor) cannot be seen as it lies beneath other muscles. It attaches to the lumbar vertebrae and the femur.

GLOSSARY

Section 1: Acronyms and Abbreviations

AC	Active Component
AGR	ability group run
AIT	advanced individual training
APFT	Army Physical Fitness Test
AR	Army regulation
ARNG	Army National Guard
ARTEP	Army Training and Evaluation Program
ATP	adenosine triphosphate
BCT	basic combat training
BDU	battle dress uniform
BPM	beats per minute
BT	basic training
BTMS	Battalion Training Management System
c	centigrade
CAD	coronary artery disease
CPR	cardiopulmonary resuscitation
CPT	captain
CR	cardiorespiratory
CVSP	cardiovascular screening program
DA	Department of the Army
DOD	Department of Defense
EDRE	emergency deployment readiness exercise
EIB	Expert Infantryman Badge
EOSB	electrically operated, stationary bicycle
F	Fahrenheit
FITT	frequency, intensity, time, type
FM	field manual
FTX	field training exercise
HDL	high-density lipoprotein
HQ	headquarters
HQDA	Headquarters, Department of the Army
HRR	heart rate reserve
ID	identification
IET	initial entry training
IG	inspector general
kph	kilometers per hour

lat	latissimus dorsi
LCE	load-carrying equipment
LDL	low-density lipoprotein
MACOM	major Army command
MEDDAC	medical department activity
METL	mission-essential task list
MFT	master fitness trainer
MHR	maximum heart rate
min	minute(s)
MOS	military occupational specialty
MPH	miles per hour
MRDA	military recommended dietary allowance
MRE	meal, ready to eat
NCO	noncommissioned officer
NCOIC	noncommissioned officer in charge
NGR	National Guard regulation
No.	number
OIC	officer in charge
OST	one-station training
OSUT	one-station unit training
Pam	pamphlet
PE	perceived exertion
PNF	proprioceptive neuromuscular facilitation
PRE	partner-resisted exercise
PT	physical training
pts	points
PU	push-up
RC	Reserve Component
rep	repetition
RHR	resting heart rate
RICE	rest, ice, compression, elevation
RM	repetition maximum
ROTC	Reserve Officers' Training Corps
sec	second(s)
SCUBA	self-contained underwater breathing apparatus
SDT	self development test
SOP	standing operating procedure
SU	sit-up
TB med	technical bulletin, medical
TDA	table of distribution and allowances
THR	training heart rate
TM	technical manual
TOE	table of organization and equipment
TRADOC	U.S. Army Training and Doctrine Command
TS	timed set
TSP	training support package

U.S.	United States
USAPFS	United States Army Physical Fitness School
USAR	United States Army Reserve
VO_2max	maximum oxygen consumption per minute
WBGTI	wet bulb globe temperature index
WCF	windchill factor

Section II: Terms

extension

An increase in the angle between two bones in which a straightening movement occurs; the opposite of flexion. For example, extension of the elbow involves an increase in the angle formed by the upper and lower arm as the arm straightens at the elbow.

flexion

A decrease in the angle between two bones in which a bending movement occurs; the opposite of extension. For example, flexion of the elbow involves a decrease in the angle formed by the lower and upper arm as the arm bends at the elbow.

REFERENCES

SOURCES USED

These are the sources quoted or paraphrased in this publication.

ARMY REGULATIONS (ARs)

15-6	Procedures for Investigating Officers and Boards of Officers. May 1988.
30-1	The Army Food Service Program. January 1985.
350-15	Army Physical Fitness Program. November 1989.
385-55	Prevention of Motor Vehicle Accidents. March 1987.

OTHER ARMY PUBLICATIONS

DOD Directive 1308.1	Physical Fitness and Weight Control Program. April 1981.
FM 21-18	Foot Marches. June 1990.

DOCUMENTS NEEDED

These documents must be available to the intended users of this publication.

ARMY REGULATIONS (ARs)

40-501	Standards of Medical Fitness. July 1987.
600-8-2	Suspension of Favorable Personnel Actions (Flags). October 1987.
600-9	The Army Weight Control Program. September 1986.
600-63	Army Health Promotion. November 1987.

OTHER ARMY PUBLICATIONS

FM 25-100	Training the Force. November 1988.
NGR 40-501	Medical Examination for Members of the Army National Guard. October 1981.
TRADOC Reg 350-6	Initial Entry Training (IET) Policies and Administration. August **1989.**

READINGS RECOMMENDED

These readings contain relevant supplemental information.

DEPARTMENT OF THE ARMY PAMPHLETS (DA Pares)

28-9	Unit Level Recreational Sports. June 1973.
350-15	Commander's Handbook on Physical Fitness. October 1982.
350-18	The Individual's Handbook on Physical Fitness. May 1983.
350-22	You and the Army Physical Fitness Test (APFT). September 1987.
351-4	Army Formal Schools Catalog. August 1991.

FIELD MANUALS (FMs)

21-150	Combative. December 1971.
22-5	Drill and Ceremonies. December 1986.
31-70	Basic Cold Weather Manual. April 1968.

OTHER ARMY PUBLICATIONS AND MATERIALS

AR 215-1	Administration of Army Morale, Welfare, and Recreation. February 1984.
DA Form 705	Army Physical Fitness Test Scorecard. May 1987.
DA Form 3349	Physical Profile. May 1986.
Folio No. 1	"Training Facilities," Corps of Engineers Drawing No. 28-13-95. Directorate of Facilities Engineering.
SB 10-260	Master Menu. December 1989.
TB Med 507	Occupational and Environmental Health Prevention, Treatment, and Control of Heat Injury. July 1980.

TSP Physical Fitness Training - Total Fitness. July 1987.

TRAINING VIDEO TAPES (TVTS)

8-103	Standards for Determining Body Fat. 1986.
21-76	Army Physical Fitness Test (APFT). 1986.
21-191	Administration of the APFT. 1988.
21-192	Partner-Resisted Exercises (PRE). 1987.
21-218	Flexibility: The Truth About Stretching. 1989.
21-203	Push-up/Sit-up Improvement. 1988.

INDEX

This is a topical index organized alphabetically. Citations are to paragraph numbers.

FM 21-20
30 SEPTEMBER 1992

By Order of the Secretary of the Army

GORDON R. SULLIVAN
General, United States Army
Chief of Staff

Official:

MILTON H. HAMILTON
Administrative Assistant to the
Secretary of the Army

02361

DISTRIBUTION:

Active Arm y, USAR and ARNG: To be distributed In accordance with DA Form 12-11E, requirements for FM 21-20, Physical Fitness Training (Qty rqr block no. 0165).

•U.S. Government Printing Office: 1994 – 300-421/82850

Made in the USA
Monee, IL
04 November 2026

67384029R00134